Aman Ghebremichael Kidanemariam

The formation of patterns in subaqueous sediment

Dissertationsreihe am Institut für Hydromechanik
Karlsruher Institut für Technologie (KIT)

Heft 2016/1

Eine Übersicht aller bisher in dieser Schriftenreihe erschienenen Bände finden Sie am Ende des Buches.

The formation of patterns in subaqueous sediment

by
Aman Ghebremichael Kidanemariam

KIT Scientific Publishing

Dissertation, Karlsruher Institut für Technologie (KIT)
Fakultät für Bauingenieur-, Geo- und Umweltwissenschaften, 2016
Tag der mündlichen Prüfung: 11. Mai 2015
Referenten: Prof. Dr. Markus Uhlmann, Prof. Dr. Élisabeth Guazzelli

Impressum

KIT Scientific Publishing

Karlsruher Institut für Technologie (KIT)
KIT Scientific Publishing
Straße am Forum 2
D-76131 Karlsruhe

KIT Scientific Publishing is a registered trademark of Karlsruhe Institute of Technology. Reprint using the book cover is not allowed.

www.ksp.kit.edu

Print on Demand 2016

ISBN 978-3-7315-0532-7
DOI 10.5445/KSP/1000054986

The formation of patterns in subaqueous sediment

Zur Erlangung des akademischen Grades eines

Doktor-Ingenieurs

von der Fakultät für

Bauingenieur-, Geo- und Umweltwissenschaften

des Karlsruher Instituts für Technologie (KIT)

genehmigte

DISSERTATION

von

Aman Ghebremichael Kidanemariam

geboren in Asmara, Eritrea

Tag der mündlichen Prüfung: 11. Mai 2015

Referent: Prof. Dr. Markus Uhlmann
Korreferent: Prof. Dr. Élisabeth Guazzelli

Karlsruhe (2016)

to Woder & Melael

Abstract

Sedimentary patterns are commonly observed in rivers and coastal waters. Providing fundamental understanding of the mechanisms involved in the formation of these patterns is highly desirable since they significantly influence flow characteristics such as resistance and sediment transport. However, the complex interaction between the sediment particles and the driving turbulent flow has made this task very challenging to the present date. This thesis aims to contribute to the elucidation of the physical mechanisms involved in the processes of subaqueous bedload transport and ripple formation. The phenomenon is numerically investigated by performing simulations which resolve all the relevant length and time scales of the flow problem. The numerical method employed features an immersed boundary technique for the treatment of the moving fluid-solid interfaces and a soft-sphere model to realistically treat inter-particle contacts.

The adequacy of the numerical approach was extensively validated with respect to two benchmark configurations. First, the problem a sphere colliding with a plane wall in a viscous fluid was considered. The simulations successfully reproduced the rebound behavior of the sphere in close agreement with available experimental measurements. Furthermore, extensive validation with respect to available experimental data of the erosion and transport of spherical particles induced by the shearing laminar flow, was performed. Beyond validation, the latter simulations have allowed us to tackle the problem of bedload transport by itself. In particular, we have addressed the scaling laws of particle flux, fluid and particle velocities inside the sediment bed, as well as the contribution of hydrodynamic and collision forces to the momentum transfer between the two phases.

The problem of pattern formation was then addressed by performing several medium- to large-scale simulations, both in the laminar and turbulent regimes. In order to realistically capture the phenomenon, sufficiently large computational domains involving up to $O(10^9)$ grid nodes were adopted, while the mobile sediment bed was represented by up to $O(10^6)$ freely moving spherical particles. Depending on the chosen value of the parameters, sediment erosion-deposition process takes place over an initially flat bed, which ultimately leads to the emergence and evolution of the patterns. The results of the simulations, with respect to the pattern wavelength, amplitude, asymmetric shape as well as their propagation velocity, are in excellent agreement with available experimental data. Furthermore, the computational box size was strategically varied in order to find a minimum box dimension which accommodates an unstable pattern wavelength, thus determining the cutoff length for pattern formation. Moreover, a rigorous dune-conditioned statistical analysis of the flow field and particle motion, which takes into account the spatial and temporal variability of the sediment bed, was performed in order to characterize the structure of the driving turbulent flow and its correlation with the bedforms. The response of particles, individually or collectively, to the flow and bedform variations was addressed as well. Of particular interest to the theoretical study of sedimentary patterns is the accurate determination of the bottom boundary shear stress along the bedforms and its relation to the local particle flow rate. The analysis addressed, among other aspects, this outstanding question, which revealed that the shear stress is maximum at a location upstream of the dune crests; hence exhibiting a positive phase shift with respect to the dune geometry. This shows that relating the local particle flux and the local shear stress by an algebraic expression, as is usually done in morphodynamic models, is not adequate.

Zusammenfassung

Unter gewissen Bedingungen kann eine kleine Störung in einem ansonsten flachen Sedimentbett eines fließenden Gewässers verstärkt werden, was zu wellenartigen (wellenformige) Sedimentformationen führen kann. Diese werden üblicherweise in der Literatur als Transportkörper (Riffel, Dünen) bezeichnet und können oftmals als Unterwasserdünen in Flussströmungen oder in Küstenregionen beobachtet werden. Ziel der vorliegenden Arbeit ist es, den der aquatischen Dünenbildung zugrundliegenden hydromechanischen Mechanismus zu untersuchen und damit zum, aus technologischer Sicht, Verständnis dieses Phänomens beizutragen. Zu diesem Zweck wurde eine Reihe von hochauflösenden numerischen Simulationen durchzufüht, in denen alle relevanten Längen- und Zeitskalen (inklusive des nahfelds um die Partikel) aufgelöst wurden. Die Ausbildung und die Entwicklung dieser Dünenformationen wurden am Beispiel von einem ebenen Kanal mit einer großen Anzahl bewegter diskreter Partikel simuliert. Eine detaillierte Analyse der numerischen Ergebnisse hat uns ermöglicht, die folgenden Fragestellungen zu untersuchen: (i) Wodurch ist das Strömungsfeld und die Partikelbewegung während der Ausbildungs- und Entwicklungsphase der Dünen charakterisiert? (ii) Wie reagieren einzelne Partikel auf die Strukturen der turbulenten Strömung? (iii) Bei welcher charakteristischen Wellenlänge wird das Sedimentbett instabil und setzt sich in Bewegung? Darüber hinaus, ermöglichen unsere Ergebnisse die Eignung bestehender Theorien zur Dünenbildung zu überprüfen und gegebenenfalls mögliche Verbesserungen vorzuschlagen. Im Folgenden werden die Ergebnisse kurz zusammengefasst.

Sedimenterosion in einer laminaren Kanalströmung

Der Transport von Partikeln als Geschiebtransport in einer laminaren Kanalströmung wurde untersucht. Dieses Teil der Arbeit ist erstens, die Eignung der vorliegenden numerischen Methode für die Simulation der Entstehung von Transportkörpern zu demonstrieren, und zweitens neue numerische Daten des Geschiebetransports zu erzeugen. Zu diesem Zweck haben wir eine große Anzahl von direkten numerischen Simulationen der Erosion einer dicken Schicht von mobilen Sedimentpartikel in einer laminaren Kanalströmung durchgeführt. Die verschiedene Simulationen unterscheiden sich im Bezug auf die Größe der Shieldszahl – der dimensionslosen Sohlenschubspannung – die die vorliegende Problemstellung charakterisiert. Unsere Ergebnisse liegen in sehr guter Übereinstimmung mit experimentellen Daten aus der Literatur und zeigen, dass die von uns benutzte numerische Methode sich für die Simulation solcher Prozesse sehr gut eignet. Insbesondere die Shieldszahl-Abhängigkeit der Sedimenttransportrate (pro Kanal-Einheitsbreite), die Dicke der beweglichen Schicht der Sohle als auch die mittlere Fluid- und Partikelgeschwindigkeiten waren sehr zufriedenstellend. Basierend auf den DNS Daten wurde nachgewiesen, dass die dimensionslose Sedimenttransportrate mit der dritten Potenz der Shieldszahl variiert, sobald der Wert der Shieldszahl den kritischen Wert für Geschiebetriebbeginn überschreittet.

Außerdem wurde gezeigt, dass die mittlere Fluidgeschwindigkeit, oberhalb des Sedimentbettes, durch ein parabelförmiges Profil charakterisiert werden kann. Die Fluid- und Partikelgeschwindigkeiten im Inneren der beweglichen Schicht der Sohle wurden vollständig durch die Definition von geeigneten Geschwindigkeit und Längenskalen beschrieben. Eine empirische Exponentialkurve, die vollständig durch die genannten Längen- und Geschwindigkeitsskalen definiert ist, wurde dann vorgeschlagen und es wurde gezeigt, dass sie die Geschwindigkeitsprofile zufriedenstellend repräsentiert. Schließlich wurde der Beitrag der hydrodynamischen und Interpartikelkollisionskräfte auf die Impulsübertragung zwischen den beiden Phasen analysiert. Der effektive Reibungskoeffizient sowie der Volumenanteil des Sedimentbettes waren durch die dimensionslose Viskosität (Boyer u. a., 2011) gut beschrieben.

Entstehung von Transportkörpern in Kanalströmung

Mehrere direkte numerische Simulationen der Entstehung von Transportkörpern über einem erodierbaren Sedimentbett sowohl im laminaren als auch im turbulenten Regime wurden durchgeführt. Die räumliche und zeitliche Variation der Sedimentbetthöhe, die aus dem durchschnittlichen Feststoffvolumenanteil bestimmt wurde, wurde analysiert, um die zeitliche Entwicklung der Eigenschaften der Transportkörper zu untersuchen. Die extrahierte Wellenlänge, die Amplitude sowie die Fortschrittgeschwindigkeit der Transportkörpern waren in sehr guter Übereinstimmung mit früheren experimentellen Beobachtungen.

Weitere Simulationen wurden durchgeführt, mit denselben physikalischen und nummerischen Parameterwerten, mit Ausnahme der Länge des Kanals. In jeder Simulation wurde die Länge des Kanals stufenweise reduziert, um die untere Grenze der instabilen Wellenlängen der wellenförmigen Muster zu bestimmen. Die Ergebnisse zeigten, dass für die in dieser Arbeit betrachteten Parameterwerte die notwendige Kanallänge für die Bildung von Dünen-Muster im Bezug auf den Durchmesser der Partikeln $78D$–$102D$ beträgt oder im Bezug auf die Halbkanalhöhe $3H$–$4H$ ist. Unterhalb dieser Länge wurde keine Dünenbildung beobachtet. Weiterhin haben wir eine Simulation in einem relativ großen Rechnengebiet mit einer grossen Anzahl bewegter diskreter Partikel durchgeführt. Es stellte sich heraus, dass die am meisten verstärkte Mode eine Wellenlänge $\lambda_c \approx 112D \approx 4{,}36H$ hat. Daraus wurde geschlossen, dass ein Rechengebiet mit $L_x/\lambda_c = O(10)$ erforderlich ist, um den natürlichen Auswahlmechanismus der Anfangswellenlänge korrekt zu erfassen.

Strömungsfeld und Partikelbewegung über den Transportkörpern

Das Strömungsfeld als auch die zugehörige Partikelbewegung über den Unterwassertransportkörpern wurde rigoros untersucht. Der Einfluss der Transportkörper auf die Eigenschaften des Strömungsfeldes wurde durch Untersuchung des Zeitverhaltens der Fluktuationen der Fluidgeschwindigkeit und der Sohlschubspannung studiert. Außerdem wurde der Einfluß der Transportkörper auf die Sedimenttransportrate untersucht. Im Fall einer laminaren Strömung ist die volumetrische Sedimenttransportrate konsistent mit der kubischen Potenz der Shieldszahl. Anderseits entspricht die Sedimenttransportrate im turbulenten Fall dem empirischen Potenzgesetz von Wong und Parker (2006), das für Erosion eines ebenen Sedimentbettes konzipiert wurde. Die hoch komplizierte Struktur der turbulenten Strömung, die auf die entstehenden Transportkörper wirkt, zeigte eine starke Korrelation mit der Topologie des Sedimentbettes. Für die Zwecke der Charakterisierung der Strömung, wurde eine dünenbedingte statistische Methode vorgeschlagen, die die räumliche und zeitliche Variabilität des Sedimentbettes berücksichtigt. Diese Methode wurde dann verwendet, um die Eigenschaften der statistisch zweidimensionalen Strömung zu studieren. In Übereinstimmung mit den Ergebnissen aus der Literatur, zeigte die turbulente kinetische Energie sowie die Reynolds Schubspannung Maximalwerte in der Rezirkulationsregion unterstrom von dem Kamm der Dünen.

In der Studie von Dünen, ist die genaue Bestimmung der räumlichen Variation der Schubspannung entlang der Dünen und ihre Beziehung zu den lokalen Sedimenttransportrate entscheidend. Zu diesem Zweck wird eine Methodik vorgestellt, die die Bestimmung der Bodenschubspannung basierend auf der Integration der mittleren Impulsgleichung ermöglicht. Auswertung der DNS-Ergebnisse zeigte, dass der Ort der maximalen Schubspannung nicht am Dünenkamm ist, sondern an einer Stelle stromaufwärts, was zu einer positiven Phasenverschiebung im Vergleich zu der Dünengeometrie führt. Es wird weiterhin angenommen, dass eine solche Phasenverschiebung die Ursache für die Instabilität eines Sedimentbettes ist (Charru u. a., 2013).

Acknowledgments

First and foremost, I would like to express my special appreciation and thanks to my supervisor Prof. Dr. Markus Uhlmann. It has been an honor to be his PhD student. I thank Markus sincerely not only for his tremendous academic support, but also for giving me so many wonderful opportunities. He has introduced me to the challenging and yet very interesting field of scientific computing applied to the problem of turbulence and sediment transport. He gave me sufficient freedom to delve myself into research and yet has cared so much about my work that he made himself always available to provided me his guidance and prompt answers to my questions. It would never have been possible for me to start this work and take it to completion without his incredible support, encouragement and supervision. Similar profound gratitude goes to Prof. Dr. Manuel García-Villalba from whom I have benefited greatly through many fruitful discussions, a lot of motivation and feedback. I would also like to thank Prof. Dr. Élisabeth Guazzelli from IUSTI, Polytech Marseille for her interest in reading and reviewing this thesis and Prof. Dr. -Ing. Peter Betsch for his disposal as a member of my PhD committee.

I am also hugely appreciative to the present and past members of the Computational Fluid Dynamics Group (a.k.a. Turbulent Flow Group) as well as the IFH stuff as a whole. Starting as a student assistant, I have been a member of this incredible group for almost eight years. The group has been a source of friendships as well as good advice and collaboration. My special thanks goes to Clemens, Todor, Fred, Agathe, Yoshi, Marco, Angelika, Conny, Helmut, Kad, Herlina, Tiago, Tim. You have substantially contributed to my personal and professional achievements.

Thanks is due to Pascale Aussillous and Élisabeth Guazzelli for sharing their experimental data in electronic form and for fruitful discussions throughout this work. I acknowledge the hospitality of the group "GEP" at IUSTI during my extended stay.

I also thank my family and friends who have provided me with moral support throughout the course of my PhD. In particular, heartfelt gratitude goes to my beloved wife; her prayer for me was what sustained me thus far. She has also rewarded me with the most precious daughter in the world. My wife and my daughter are the most important people in my world and I dedicate this thesis to them.

This work was supported by the German Research Foundation (DFG) through Grant UH 242/2-1 for which I am grateful. All the computations have been carried out at the Leibniz Supercomputing Center (LRZ) of the Bavarian Academy of Science and Humanities (Grants `pr58do` and `pr84du`). I thankfully acknowledge, the computer resources, technical expertise and assistance provided by the staff at LRZ München. Moreover, I am grateful for the internal funding I received from KIT under the funding instrument 'Feasibility Studies of Young Scientists'. The funding gave me the opportunity to demonstrate the feasibility of performing direct numerical simulation of pattern formation.

Related publications

The content of this thesis is based in parts on the following references, published during the course of this PhD thesis. The contributions of the co-authors is highly acknowledged.

Peer-reviewed publications

García-Villalba, M., A. G. Kidanemariam, and M. Uhlmann (2012). "DNS of vertical plane channel flow with finite-size particles: Voronoi analysis, acceleration statistics and particle-conditioned averaging". *Int. J. Multiph. Flow*, 46, pp. 54–74.

Kidanemariam, A. G., C. Chan-Braun, T. Doychev, and M. Uhlmann (2013). "Direct numerical simulation of horizontal open channel flow with finite-size, heavy particles at low solid volume fraction". *New J. Phys.* 15(2), p. 025031.

Kidanemariam, A. G. and M. Uhlmann (2014a). "Direct numerical simulation of pattern formation in subaqueous sediment". *J. Fluid Mech.* 750, R2.

Kidanemariam, A. G. and M. Uhlmann (2014b). "Interface-resolved direct numerical simulation of the erosion of a sediment bed sheared by laminar channel flow". *Int. J. Multiph. Flow*, 67, pp. 174–188.

Conference contributions

García-Villalba, M., A. G. Kidanemariam, and M. Uhlmann (2012). "Some aspects of the interaction between turbulent flow and fintite-size particles". Nonlinear transport, dynamics and fluctuations in condensed matter physics, Madrid (Spain).

Kidanemariam, A. G., C. Chan-Braun, and M. Uhlmann (2012). "Numerical simulation of horizontal turbulent flow with heavy finite-size particles". 9. EUROMECH European Fluid Dynamics Conference, Rome (Italy).

Kidanemariam, A. G. and M. Uhlmann (2014). "DNS of the initiation of subaqueous sedimentary patterns in a horizontal channel flow". 9. Euromech European Fluid Mechanics Conference, Copenhagen (Denmark).

Uhlmann, M., T. Doychev, A. G. Kidanemariam, and A. Chouippe (2014). "The immersed boundary method as a tool for studying turbulence-particle interactions". ICNMMF-II, Darmstadt (Germany).

Contents

Nomenclature

Mathematical symbols

$*^+$	quantity $*$ normalized in wall units u_τ and δ_ν
$*'$	fluctuation of a quantity $*$ with respect to its mean $\langle * \rangle$
$*''$	dune-form induced fluctuation of a quantity $*$
$\langle \cdot \rangle_D$	averaging operator over a given number of dunes at a given time
$\langle \cdot \rangle_i$	averaging operator in the i-th direction
$\langle \cdot \rangle_{ij}$	averaging operator in the i-th and j-th directions
$\langle \cdot \rangle_{it}$	averaging operator in the i-th direction and in time
$\langle \cdot \rangle_t$	averaging operator in time
$\langle \cdot \rangle_{xyz}$	averaging operator in all spatial directions
$\langle \cdot \rangle$	averaging operator in time and in all homogeneous directions
∇^2	Laplace operator, $\nabla^2 = (\partial_x^2, \partial_y^2, \partial_z^2)^T$
∇	Nabla operator, $\nabla = (\partial_x, \partial_y, \partial_z)^T$
$\widehat{*}$	quantify $*$ as observed in a moving inertial frame of reference
$\widetilde{*}$	phase shifted quantify of $*$
$\mathrm{d}/\mathrm{d}i$	derivative in i direction
$\partial/\partial i$, ∂_i	partial derivative in i direction

Greek symbols

ΔU	roughness function, i.e. velocity offset due to roughness in law of the wall
Δ_c	parameter of the collision model; force range between contacting particle surfaces
Δ_x	streamwise width of discrete streamwise-orthogonal bins for averaging
Δt_{sub}	time step adopted for integration of the Newton equations of motion
Δt	time step adopted for integration of the Navier-Stokes equations
Φ_s	global solid volume fraction
Φ_{bed}	average solid volume fraction of a sediment bed

Θ_c	critical value of Shields number for incipient sediment motion
Θ_{Pois}	Shields number defined under the assumption of a laminar Poiseuille flow
Θ	Shields number
ω_p	angular velocity of particle with respect to its center of mass $\boldsymbol{\omega}_p = (\omega_{px},\omega_{py},\omega_{pz})^T$
ω	vorticity $\boldsymbol{\omega} = (\omega_x,\omega_y,\omega_z)^T$
τ_f	fluid shear stress tensor $\boldsymbol{\tau}_f = \nu\rho_f(\partial_j u_{fi} + \partial_i u_{fj})$
τ_p	particle shear stress tensor stemming from the fluid-particle interaction
δ_v	viscous length scale
$\delta_c^{(i,j)}$	parameter of the collision model; penetration (overlap) length between particles i and j
δ_{ij}	Kronecker delta
δt	time separation
δx	streamwise distance separation
$\dot{\gamma}_f$	fluid phase shear rate
$\dot{\gamma}_p$	particle phase shear rate
λ_c	critical unstable wavelength of a sediment bed
λ_h	average bedform wavelength
λ_i	wavelength of the i-th harmonic in the wavenumber space
λ_{th}	lower threshold of the unstable wavelength of a sediment bed
μ_c	parameter of the collision model; Coulomb friction coefficient
μ_{eff}	effective coefficient of friction of a sediment bed
ν	kinematic viscosity of fluid
$\overline{\overline{\sigma}}$	fluid stress tensor
ϕ_f	fluid phase indicator function
ϕ_p	solid phase indicator function
ϕ_s	solid volume fraction compute from particle number density in bins
ρ_f	density of fluid
ρ_p	density of particles
σ_h	r.m.s. fluctuation of a sediment bed height
σ_j	growth rate of the amplitude of the j-th harmonic
τ^W	fluid shear stress at the wall of a channel flow
τ_{bot}	dune boundary shear stress
τ_{tot}	total shear stress
ε_d	parameter of the collision model; dry restitution coefficient

ε	effective coefficient of restitution
φ_j	phase of the j-th harmonic
h_m	thickness of the mobile sediment layer
$\Delta x, \Delta y, \Delta z$	computational grid spacing

Roman symbols

AR	aspect ratio of a dune geometry
AR	degree of asymmetry of a dune geometry
A_h	difference in wall-normal coordinate of the local maximum and that of adjacent local minimum of a given dune
A_j	amplitude of the j-th harmonic
D	diameter of a spherical particles
F_W	submerged weight of a particle
Ga	Galileo number
H	half-height of channel flow (full channel height in the case of open channel flow)
I_v	dimensionless viscous number
$I_{stat}^{(l)}$	an indicator function which has a value of unity if the l-th particle is stationary
L_i	computational domain length in the i-th direction
L_{sat}	saturation length scale
M_i	mass of a particle with index i
$M_{i,j}$	reduced mass of a particle pair i,j
N_p	total number of spherical particles in a given simulation
P^p	confining pressure exerted on a submerged sediment bed
R_h	two-point autocorrelation function of a sediment bed height fluctuation
$R_{h\tau}$	cross-correlation function between the sediment bed height and the boundary shear stress
R_{ht}	space-time correlation function of a sediment bed height fluctuation
Re_τ	friction velocity based Reynolds number
Re_b	bulk velocity based Reynolds number
Re_p	Reynolds number based on particle diameter and its characteristic velocity
Re	Reynolds number
R	radius of a spherical particles
S_p	enclosing surface of a rigid particle
St	Stokes number

T_b	bulk time unit
T_c	parameter of the collision model; collision duration
T_{obs}	simulation observation interval over which statistics are collected
U_f^0	value of the mean particle velocity at the fluid-bed interface y_0
U_f^m	maximum value of the mean velocity profile of a laminar channel flow
U_g	gravitational velocity scale
U_s	Stokes settling velocity
V_p	volume of a particle; for a spherical particle $V_p = \pi D^3/6$
$\boldsymbol{F}_d^C$	damping collision force acting normal to the surface of contacting particles
$\boldsymbol{F}_{el}^C$	elastic repulsive collision force acting normal to the surface of contacting particles
$\boldsymbol{F}_t^C$	damping collision force acting tangential to the surface of contacting particles
$\boldsymbol{F}^C$	total collision force acting on a given particle $\boldsymbol{F}^C = (F_x^C, F_y^C, F_z^C)^T$
$\boldsymbol{F}^H$	total hydrodynamic force acting on a given particle $\boldsymbol{F}^H = (F_x^H, F_y^H, F_z^H)^T$
$\boldsymbol{F}$	total force acting on a given particle $\boldsymbol{F} = (F_x, F_y, F_z)^T$
$\boldsymbol{I}_p$	moment of inertia about the center of mass of a particle
$\boldsymbol{T}^C$	collision torque acting on particles $\boldsymbol{T}^C = (T_x^C, T_y^C, T_z^C)^T$
$\boldsymbol{T}^H$	hydrodynamic torque acting on particles $\boldsymbol{T}^H = (T_x^H, T_y^H, T_z^H)^T$
$\boldsymbol{T}$	total torque acting on particles $\boldsymbol{T} = (T_x, T_y, T_z)^T$
$\boldsymbol{X}_p$	Lagrangian position vector of the center of mass of a particle
$\boldsymbol{X}$	Lagrangian position vector $\boldsymbol{x} = (x,y,z)^T$
$\boldsymbol{f}$	Eulerian volume force acting on the fluid at the fluid-particle interface $\boldsymbol{f} = (f_x, f_y, f_z)^T$
$\boldsymbol{g}$	vector of gravitational acceleration
$\boldsymbol{n}$	unit normal vector
$\boldsymbol{u}_f$	velocity vector of fluid $\boldsymbol{u}_f = (u_f, v_f, w_f)^T$
$\boldsymbol{u}_p$	translational velocity vector of particle $\boldsymbol{u}_p = (u_p, v_p, w_p)^T$
$\boldsymbol{u}_r^{i,j}$	relative velocity of particle i with respect to particle j
$\boldsymbol{x}_p$	Eulerian position vector of the center of mass of a particle
$\boldsymbol{x}$	Eulerian position vector $\boldsymbol{x} = (x,y,z)^T$
$\hat{A}_h$	single-sided amplitude spectrum
$\hat{h}_{bj}$	DFT of the bed height fluctuation
$\mathcal{L}$	characteristic lengthscale
$\mathcal{U}$	characteristic velocity scale
$\mathcal{V}$	control volume

ϕ_m	maximum value of solid volume fraction of a given sediment bed
$\tilde{h}$	characteristic half thickness of the mobile bed
c_j	phase speed of the j-th harmonic
c_{dn}	parameter of the collision model; normal damping coefficient
c_{dt}	parameter of the collision model; tangential damping coefficient
h^*	distance of the location of U_f^m measured from the top wall of a laminar channel flow
h_b	total sediment bed height
h_f	fluid height
k_n	parameter of the collision model; stiffness coefficient
k	turbulent kinetic energy
p	fluid pressure
q_f	volumetric flowrate of fluid phase (per unit span)
q_i	inertial flowrate scale
q_p	volumetric flowrate of solid phase (per unit span)
q_v	viscous flowrate scale
q_{h_f}	flowrate scale similar to q_v but based on clear fluid height rather than particle diameter
r_x^m	streamwise location of the maximum value of R_{ht}
u_D	mean dune propagation velocity
u_τ	friction velocity
u_b	bulk velocity
v_{pT}	terminal settling velocity of a particle
x,y,z,t	Cartesian coordinates (streamwise, wall-normal and spanwise directions) and time
y_0	wall-normal coordinate of the fluid-bed interface location

Abbreviations

CPU	central processing unit
DEM	discrete element modelling/method
DFT	discrete Fourier transformation
DNS	direct numerical simulation
IBM	immersed boundary method
LES	large eddy simulation
RANS	Reynolds averaged Navier-Stokes equations
r.m.s.	root mean square

Chapter 1

Introduction

1.1 Background

When a body of fluid (water, air, etc.) flows over a sediment bed, erosion may take place which over time can lead to a change in the morphology of the bed itself. In this manner, wind, rivers and ocean currents have been shaping the face of the earth for many years. Analogously, fluid flow through man-made canals and technical devices has often the consequence of inducing a transport of sediment particles of some sort. Once the motion of sediments is initiated, a complex interplay between the particles and the carrier fluid ensues, especially when the base flow is *turbulent*. Turbulence means that the boundary layer developing near the sediment bed is populated by a large number of transient flow structures (i.e. turbulent eddies), which are effective to varying degrees in lifting-up particles from the bed. Similarly, re-deposition of particles is controlled by the instantaneous turbulent flow structures.

The process of local erosion of particles from an erodible sediment bed and their deposition at certain preferential locations often leads to the formation of geometrically regular, repeating features, commonly termed as *sedimentary patterns* or *bedforms*[1]. Sedimentary patterns are observed in a variety of environments such as deserts, river and marine flows as well as various technical applications involving shear flow over a bed of mobile sediment particles (cf. figure 1.1). They cover a wide range of spatial and temporal scales.

From an environmental or an engineering point of view it is highly desirable to be able to predict the occurrence and the nature of the formation of patterns, since bedforms significantly influence flow characteristics such as resistance, mixing properties and sediment transport. These modified flow quantities in turn have far-reaching consequences for the operation of diverse facilities in hydraulics, chemical engineering, etc. (Yalin, 1977; Best, 2005; García, 2008b; Coleman and Nikora, 2011).

The field of sedimentary patterns is a vast subject. In the following, we give an overview of the different classification of bedforms, at the same time mentioning in which classification the present work is focused on. Bedforms could be broadly classified into *aeolian* and *subaqueous*, based on the medium under which they occur. The former being driven by wind forces while the latter form under water or a similar fluid. Although the basic formation process is the same in both categories, there are fundamental differences regarding the relevant

[1] The terminologies 'bedforms', 'sedimentary patterns' or simply 'patterns' are used interchangeably throughout the thesis and bear the same meaning.

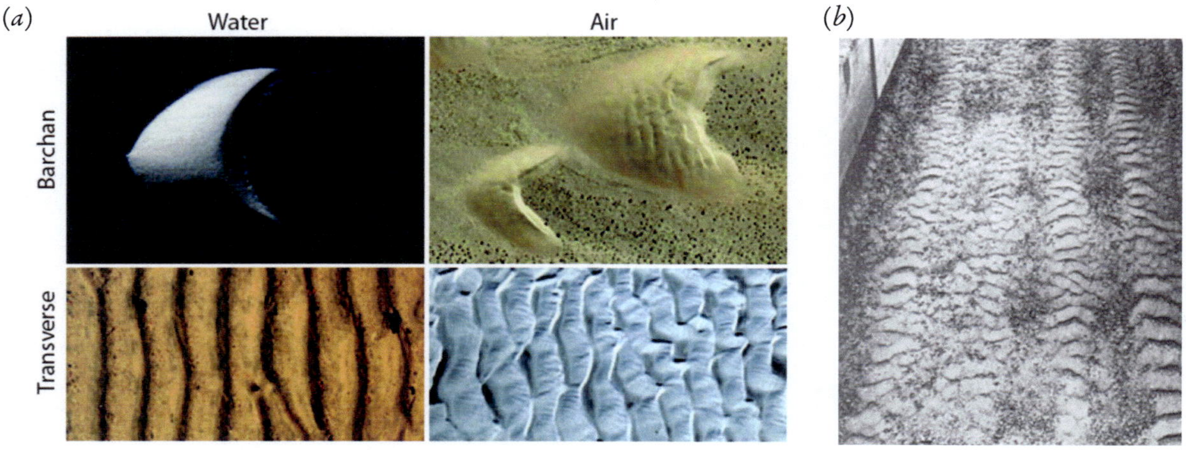

Figure 1.1: Images of sedimentary patterns. (*a*) three-dimensional barchan dunes and quasi-two-dimensional transverse dunes forming both under water and air (Charru et al., 2013). (*b*) ripples forming in between streamwise aligned longitudinal particle streaks (Seminara, 2010).

controlling mechanisms of their generation. For instance, the ratio between the density of the sediment grains and that of the carrier fluid is much smaller in the case of subaqueous bedforms when compared to that of aeolian ones. Consequently, the way sediment particles respond to fluid mediations substantially differs. The present work is restricted to the study of subaqueous bedforms, and thus is the subsequent literature review. For a detailed account of the aeolian bedforms, the reader should refer to Bagnold (1941, 1956); Sauermann et al. (2001); Andreotti et al. (2002); Andreotti and Claudin (2013) and references therein.

Subaqueous bedforms could be further categorized into those that form under the action of a *steady* flow and those that are worked by an unsteady *oscillatory* flow. By steady, it is meant that mean flow properties, such as the velocity of the fluid, do not vary with time at locations sufficiently far away from the bedforms. On the contrary, in unsteady flows, mean flow properties are functions of time. Bedforms that are observed in river beds and man-made canals are typical examples of patterns in steady flows, while those which are generated by sea wave motion in coastal regions fall under the category of bedforms which are induced by oscillatory flow. In the present work, we further restrict our study by considering bedforms occurring only in steady flows. An interested reader can refer to, for instance, Sleath (1976); Blondeaux (1990); Vittori and Blondeaux (1990); Scherer et al. (1999); Blondeaux (2001); Rousseaux et al. (2004); Charru and Hinch (2006b); Blondeaux et al. (2015) for an in-depth investigation of bedforms in oscillatory flows.

Strictly speaking, subaqueous bedforms, even those forming under steady flow conditions, are three-dimensional, exhibiting undulations which are function of both the streamwise and cross-stream directions. Nevertheless, under certain circumstances, bedforms tend to be statistically invariant in the cross-stream direction and can be considered as essentially two-dimensional (Yalin, 1977). Indeed, a large part of the previous theoretical modeling, experimental measurement as well as field observation on subaqueous bedforms, has been based on the assumption that bedforms, as well as the flow over them, are two-dimensional (Yalin, 1977; Best, 2005). In natural river channels on the other hand, it is common to observe bedforms which are highly three-dimensional affecting the mean flow in all spatial directions (Allen, 1968; Best, 2005). For such bedforms, it is not sufficient to consider irregularities only in the streamwise direction. A particular type of three-dimensional subaqueous bedforms are the *Barchan dunes* which are isolated features having a distinct crescent like shape (Endo et al., 2005; Groh et al., 2008; Franklin and Charru, 2009). The study of such three-dimensional dunes, although equally important and fascinating as their two-dimensional counterparts, is outside the scope of the present work. The thesis is mainly focused to the study of transverse bedforms, and the associated fluid flow, which are statistically invariant in the cross-stream direction.

So far, we have systematically narrowed our focus to the statistically two-dimensional subaqueous patterns which form under steady flow conditions[2]. Even at this level, it is a common practice to classify bedforms based upon their dimensions, shape, resistance to flow, mode of sediment transport, etc (Julien, 1998). Classically, a distinction is made among *ripples*, *dunes* and *antidunes* (Engelund and Fredsoe, 1982; García, 2008b). Ripples and dunes are similar. They exhibit an asymmetric triangular-like shape with a gentle slope on the upstream side of their crest and a steeper slope on the downstream side. The principal distinction between ripples and dunes is the separation of length scales; the wavelength of observed ripples is commonly believed to scale with the grain size, and that of larger scale dunes is supposedly controlled by the flow geometry, typically the flow depth (Yalin, 1977). The schematic diagram in figure 1.2 shows the different formation stages of bedforms with increasing stream power[3]. Experiments shows that at low flow rates small-scale undulations appear out of an initially flat sediment bed, having no clear-cut shape (Coleman and Melville, 1994). When increasing the flow rate, these ripples give way to the much larger dunes, which are characterized by flow separation and recirculation in their downstream faces, and which propagate downstream. At even higher flow rates antidunes appear, which tend to propagate upstream. In the case of a flow with a free surface, both dunes and antidunes dynamically interact with the surface waves. It is believed that the free surface influences the characteristics of turbulent flow over such dunes (see e.g. Best, 2005). Nevertheless, it should be mentioned that bedforms are equally observed in closed conduits which possess no free surface, and thus the existence of a free surface is not necessary for their formation (Yalin, 1977; Bridge and Best, 1988; Langlois and Valance, 2007; Ouriemi et al., 2009a). The problem of bedform and free-surface interaction is not within the scope of the present study.

Finally, let us further narrow the focus of thesis based on the 'state' of bedforms. Studies on sedimentary patterns can be broadly categorized into those which are focused on the bedform initiation and development processes and those which deal with aspects of the equilibrium configuration where bedforms are developed (ASCE Task Committee, 2002). The majority of previous experimental and field studies fall under the latter category. This is mainly due to the relative ease of experimentally measuring 'unchanging' bedform characteristics. On the other hand, studies on bedform generation and development are very limited, mainly as a result of challenges encountered in measuring or modeling them (ASCE Task Committee, 2002). The present contribution is concerned with the initiation and evolution aspects of bedforms.

1.2 Research objectives

The objective of the thesis is to investigate the formation of sedimentary patterns by means of high-fidelity numerical simulations which resolve all the relevant length and time scales of the turbulent flow, even the near-field around the individual particles. We have simulated the formation of these patterns in a channel flow configuration with a very large number of freely moving spherical particles representing the mobile bed. Such highly resolved numerical simulation is required to, first and foremost, provide a spatially and temporally resolved numerical data which contains a complete information of the flow field as well as

---

[2] Hereafter, by 'bedforms', it is meant 'two-dimensional subaqueous bedforms under steady flow conditions'

[3] Stream power is the rate of energy dissipation against the bed and banks of a river or stream per unit downstream length

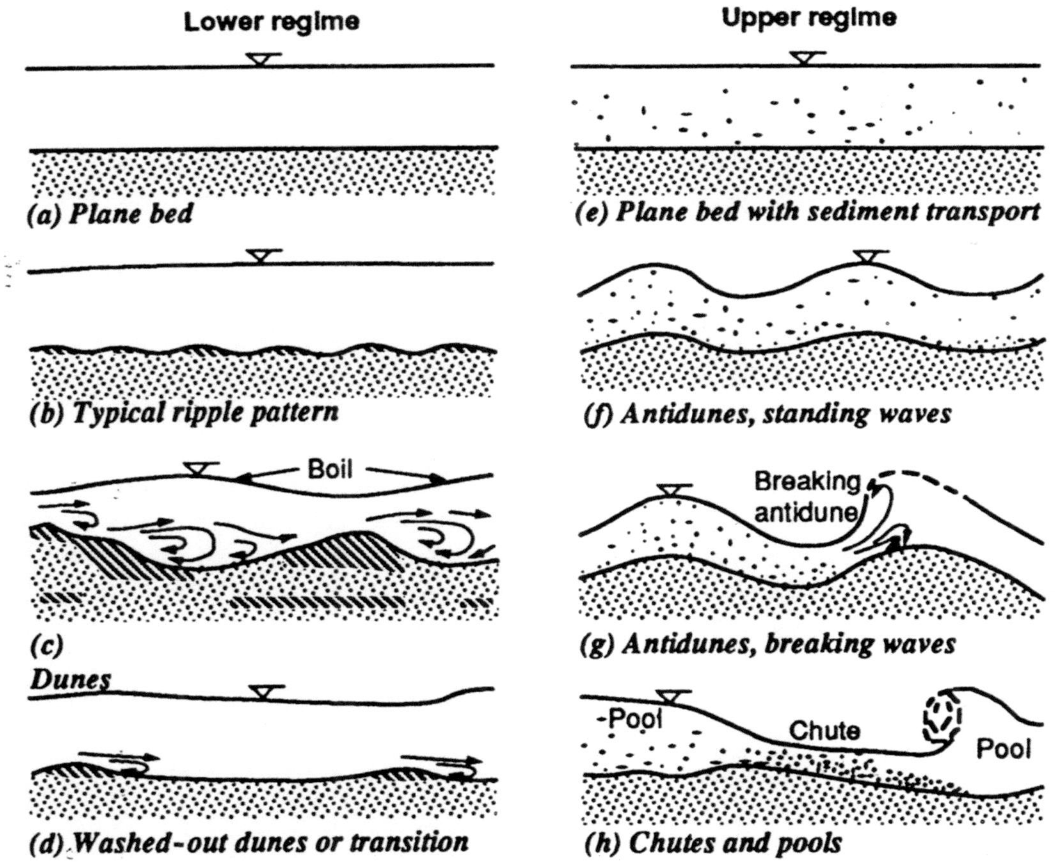

Figure 1.2: Typical bedforms in order of increasing fluid bottom friction (Engelund and Fredsoe, 1982; Julien, 1998).

individual particle-related quantities. Secondly, detailed analysis of the generated data allows us to address the different mechanisms involved in the processes of pattern formation during the initial stages of bed instability and subsequent bedform evolution. Moreover, we are in a position to test various existing models for pattern formation based upon generated data for the purpose of verifying or falsifying certain hypotheses.

1.3 Outline of the thesis

The thesis is organized as follows: In chapter 2, an overview of the fundamental concepts which are relevant for the present study is provided. Furthermore, an extensive literature review of the state-of-the-art of the initiation and evolution aspects of sedimentary patterns is presented. In chapter 3, a detailed description of the employed numerical method as well as related validation studies performed is presented. Chapter 4 presents the the detailed numerical simulation and analysis performed to study the problem of sediment erosion sheared by a laminar flow. Chapter 5 presents the details of the computational setup as well as the results of the extensive data analysis performed for of the formation of patterns. Chapter 6 is devoted to the study of the flow field and particle motion over the evolving bedforms. Concluding remarks are made in chapter 7. Some additional material is provided in the appendices.

Chapter 2

Literature review

2.1 Fundamental concepts

Some fundamental concepts are briefly recalled in this section. The aim is not to provide a detailed review, rather to familiarize the reader with the fundamental background physics which is relevant in the present thesis.

2.1.1 Governing equations of fluid and particle motion

The motion of a constant property Newtonian fluid

Let us consider the flow of an incompressible Newtonian fluid, with a constant density ρ_f, and constant kinematic viscosity ν. For a Newtonian fluid, the surface forces acting on a fluid element are given by the stress tensor $\overline{\overline{\sigma}}$ which has components

$$\sigma_{ij} = -p\delta_{ij} + \rho_f \nu \left(\partial_j u_{fi} + \partial_i u_{fj}\right) \qquad i,j = 1,2,3 \tag{2.1}$$

where u_{fi} is the velocity of the fluid in the ith direction, p is the pressure and δ_{ij} is the Kronecker delta[1]. The motion of the fluid is governed by the Navier-Stokes equations which stem from the conservation of mass and momentum. In a non-conservative form, they can be written as

$$\nabla \cdot \boldsymbol{u}_f = 0 \tag{2.2a}$$

$$\frac{\partial \boldsymbol{u}_f}{\partial t} + \left(\boldsymbol{u}_f \cdot \nabla\right) \boldsymbol{u}_f = -\frac{1}{\rho_f}\nabla p + \nu \nabla^2 \boldsymbol{u}_f + \frac{1}{\rho_f}\boldsymbol{f} \tag{2.2b}$$

where $\boldsymbol{u}_f(\boldsymbol{x},t)$ and $p(\boldsymbol{x},t)$ are the fluid velocity and pressure at position $\boldsymbol{x}$ and time t, respectively. $\boldsymbol{f}(\boldsymbol{x},t)$ represents any external body force per unit volume acting on the fluid, for instance gravity. The action of the uniform gravitational body force can be expressed in the form of a surface integral of a potential energy per unit mass, and customarily, it is absorbed into the pressure term (Batchelor, 1967; Pope, 2000). In such case, p is a modified pressure and $\boldsymbol{f}$ does not include gravity force. Due to the fact that the fluid is incompressible,

[1] Kronecker delta $\delta_{ij} = 1$ if $i = j$ and zero else.

pressure is not determined from density and temperature by an equation of a state, rather it is coupled with the velocity field and takes whatever value to enforce a divergence-free velocity field. To this end, a Poisson equation is formulated for the pressure by taking the divergence of the Navier-Stokes equations:

$$\nabla^2 p = -\rho_f \frac{\partial u_{fi}}{\partial x_j} \frac{\partial u_{fj}}{\partial x_i} \, . \tag{2.3}$$

Satisfying this equation is necessary and sufficient to insure divergence-free velocity field (Pope, 2000).

Boundary conditions

In order to completely describe a particular flow, appropriate initial and boundary conditions must be imposed. If one defines $\boldsymbol{n}$ to be the unit normal vector at an interface representing the boundary which is moving with a velocity $\boldsymbol{u}_b$, the *impermeability* condition

$$\boldsymbol{n} \cdot (\boldsymbol{u}_f - \boldsymbol{u}_b) = 0 \tag{2.4}$$

is imposed when there is no mass flux across the boundary. For viscous fluids, the fluid velocity at all fluid-solid boundaries is equal to that of the solid boundary. To satisfy this condition, the *no-slip* condition is imposed:

$$\boldsymbol{n} \times (\boldsymbol{u}_f - \boldsymbol{u}_b) = 0 \, . \tag{2.5}$$

The top boundary of open channel flows is usually modeled by the *free-slip* condition

$$u_n = 0 \quad \text{and} \quad \frac{\partial u_t}{\partial n} = 0 \tag{2.6}$$

where u_t and u_n are defined as the tangential and normal components of the velocity vector at the interface. When representing a free-surface boundary of a given flow with the free-slip condition, interface deformations at boundary are not taken into account and are assumed to be negligible. A Neumann boundary condition is imposed for the elliptic Poisson pressure equation (2.3) viz.

$$\frac{\partial p}{\partial n} = cnst. \tag{2.7}$$

Reynolds number similarity

Under the choice of appropriate characteristic length and velocity scales of a given flow, denoted as $\mathcal{L}$ and $\mathcal{U}$ respectively, the set of equations governing the motion of incompressible fluid can be written in a dimensionless form which are described by a single non-dimensional number, namely the Reynolds number

$$Re = \frac{\mathcal{U}\mathcal{L}}{\nu} \, . \tag{2.8}$$

The Reynolds number gives a measure of the ratio of inertia forces ($\rho_f \mathcal{U}^2 \mathcal{L}^2$) to the viscous forces ($\rho_f \nu \mathcal{U} \mathcal{L}$). For very small values of the Reynolds number ($Re \ll 1$), viscous forces are dominant and the flow is commonly found to be laminar. On the other hand, for very large values of the Reynolds number ($Re \gg 1$), the inertial terms dominate and the flow is turbulent. For moderate values of Re, both forces contribute and the flow is said to be in the transitional regime.

The motion of an object immersed in a Newtonian fluid

Let us consider a rigid, homogeneous impermeable object fully submerged in a constant property Newtonian fluid. The object has a density ρ_p, a volume V_p and is enclosed by a surface S_p. The motion of the object is fully described by Newton-Euler equations of linear and angular momentum of a rigid body and the Navier-Stokes equations, both coupled through the no-slip boundary condition imposed at the surface S_p.

Denoting $\boldsymbol{X}_p(t,\boldsymbol{X}_0)$ as the Lagrangian position vector, with reference to a given coordinate system, of the center of mass of the object at time t, $\boldsymbol{X}_0$ being its position at a reference time t_0, the Lagrangian velocity $\boldsymbol{U}_p(t,\boldsymbol{X}_0)$ of the object's center of mass is given by (Pope, 2000)

$$\boldsymbol{U}_p(t,\boldsymbol{X}_0) = \frac{\partial \boldsymbol{X}_p}{\partial t} = \boldsymbol{u}_p(\boldsymbol{X}_p,t)\,, \tag{2.9}$$

where $\boldsymbol{u}_p$ is the Eulerian particle velocity. Integrating equation (2.9) yields the trajectory as a function of time of the object's center of mass.

Newton-Euler equation of linear momentum states (see e.g. Jackson, 2000):

$$\rho_p V_p \frac{\mathrm{d}\boldsymbol{u}_p}{\mathrm{d}t} = \oint_{S_p} \overline{\overline{\sigma}} \cdot \boldsymbol{n} \mathrm{d}S + (\rho_p - \rho_f)V_p \mathbf{g} + \sum_{i=1}^{N_{col}} \boldsymbol{F}_i^C \tag{2.10}$$

where $\mathbf{g}$ is vector of gravitational acceleration and $\boldsymbol{n}$ is an outward pointing normal vector on the surface S_p. In addition to hydrodynamic surfaces forces and gravitational body force, the object could experience additional force as a result of collision with neighboring submerged object(s) and/or with solid boundary. In (2.10) it is assumed that the object experiences N_{col} collisions on its surface at a given instant and $\boldsymbol{F}_i^C$ represents the resultant collision force acting on the object at the ith contact location. Similarly, the Newton-Euler equation for angular momentum states (see e.g. Jackson, 2000):

$$\frac{\mathrm{d}}{\mathrm{d}t}\left(\boldsymbol{I}_p \cdot \omega_p\right) = \oint_{S_p} \boldsymbol{r} \times \left(\overline{\overline{\sigma}} \cdot \boldsymbol{n}\right) \mathrm{d}S + \sum_{i=1}^{N_{col}} \left(\boldsymbol{r}_i \times \boldsymbol{F}_i^C\right) \tag{2.11}$$

where $\boldsymbol{I}_p$ and ω_p are the moment of inertia tensor about the center of mass and the angular velocity of the object. $\boldsymbol{r} \equiv \boldsymbol{X} - \boldsymbol{X}_p$ is the position vector of the surface, relative to the center of mass of the object. For a spherical object, due to symmetry, the moment of intertia tensor is in principal axis form and each principal moment is identical. Thus, the LHS of equation (2.11) simply becomes $I_p \mathrm{d}\omega_\mathrm{p}/\mathrm{d}t$.

2.1.2 Turbulence in open channel flows

Majority of sediment-laden flows in open channels and natural streams are turbulent, characterized by unsteady, nonuniform and random motion of fluid particles. It is convenient to decompose the highly irregular motion of a turbulent flow into a mean and a fluctuating components. Such a separation is called the Reynolds decomposition. For instance the random fluid velocity field can be decomposed as

$$\boldsymbol{u}_f(\boldsymbol{x},t) = \langle \boldsymbol{u}_f(\boldsymbol{x},t) \rangle + \boldsymbol{u}'_f(\boldsymbol{x},t)\,, \tag{2.12}$$

where the mean of the random field is defined as

$$\langle \boldsymbol{u}_f(\boldsymbol{x},t)\rangle = \int_{-\infty}^{\infty} \boldsymbol{V} f(\boldsymbol{V};\boldsymbol{x},t)\mathrm{d}\boldsymbol{V} \ , \tag{2.13}$$

where $f(\boldsymbol{V};\boldsymbol{x},t)$ is the joint probability density function of the random variable $\boldsymbol{V}(\boldsymbol{x},t)$ (Pope, 2000). With this definition, the 'Reynolds averaged Navier-Stokes equations', or simply RANS along with the continuity equation, become

$$\nabla \cdot \langle \boldsymbol{u}_f\rangle = 0 \tag{2.14a}$$

$$\frac{\partial \langle \boldsymbol{u}_f\rangle}{\partial t} + \left(\langle \boldsymbol{u}_f\rangle \cdot \nabla\right)\langle \boldsymbol{u}_f\rangle = -\frac{1}{\rho_f}\nabla\langle p\rangle + \nabla\left[\nu\nabla\langle \boldsymbol{u}_f\rangle - \langle \boldsymbol{u}'_f \otimes \boldsymbol{u}'_f\rangle\right] , \tag{2.14b}$$

where $\langle \boldsymbol{u}'_f \otimes \boldsymbol{u}'_f\rangle$[2] is the Reynolds stress tensor which represents the mean momentum flux due to the fluctuating velocity field (Pope, 2000).

Now, let us consider a fully developed, statistically stationary smooth wall open channel flow. In such a flow, statistics are invariant in the streamwise and spanwise directions as well as in time. Defining a coordinate system $\boldsymbol{x} = (x,y,z)^T$ which is aligned in the streamwise, wall-normal and spanwise direction respectively, and the components of the velocity vector in the coordinate system as $\boldsymbol{u}_f = (u_f,v_f,w_f)^T$, the streamwise mean momentum equation can be reduced to

$$\frac{\mathrm{d}\langle p\rangle}{\mathrm{d}x} = \frac{\mathrm{d}\tau_f}{\mathrm{d}y} \tag{2.15}$$

where τ_f is the total shear stress given by the sum of the viscous and Reynolds shear stresses as

$$\tau_f = \rho_f \nu \frac{\mathrm{d}\langle u_f\rangle}{\mathrm{d}y} - \rho_f\langle u'_f v'_f\rangle \ . \tag{2.16}$$

That is, the mean axial pressure gradient is constant across the flow and it is balanced by the shear stress gradient. From (2.17), it follows that

$$\tau_f(y) = \tau_w\left(1 - \frac{y}{H}\right) \tag{2.17}$$

where H is the thickness of the channel and τ_w is the shear stress at the wall. In channel flows, length and velocity scales are defined based on viscosity and the wall shear stress which describe the flow very near the wall. These are the friction velocity

$$u_\tau = \sqrt{\tau_w/\rho_f} \tag{2.18}$$

and the viscous lengthscale

$$\delta_v = \nu\sqrt{\rho_f/\tau_w} = \nu/u_\tau \ . \tag{2.19}$$

Throughout the thesis, any quantity which is normalized by δ_v and/or u_τ is said to be expressed in terms of *wall units* and is superscripted by the '+' sign.

[2] the diadic product of two vectors $\boldsymbol{a}$ and $\boldsymbol{b}$ is a tensor given by $\boldsymbol{a} \otimes \boldsymbol{b} = a_i b_j$.

2.1.3 Morphodynamics – Sediment continuity equation

The motion of individual sediment grains in a given river or canal and the resulting evolution of the sediment bed is fully described by the Navier-Stokes and the Newton equations. Nevertheless, solving these equation poses immense challenge due to their complexity and dependence on multitude of parameters. Thus it has been a common practice to address the complex problem by resorting to simplified models which, to within the considered assumptions, provide a reasonable approximation to the problem. In such a spirit, in morphodynamics[3], the temporal evolution of a sediment bed $h_b(\boldsymbol{x},t)$ is related to the spatial variation of the local sediment volumetric flow rate $\boldsymbol{q}_p(\boldsymbol{x},t)$ through the sediment continuity equation model (see for instance Seminara, 2010; Charru et al., 2013):

$$\Phi_{bed}\frac{\partial h_b}{\partial t} + \nabla \cdot \boldsymbol{q}_p = 0 \tag{2.20}$$

where Φ_{bed} is the solid volume fraction of the underlying sediment bed. Equation (2.20) is classically known as the Exner equation, after Felix Maria Exner who was the first to formulated it (Exner, 1925). Usually, the sediment flow rate is expressed in terms of a suitable parameter of the flow, usually the bottom shear stress, which provides a means of coupling the sediment continuity equation with the (reduced) equations of fluid motion. The coupled formulation is then used to predict flow-induced evolution of a given sediment bed. Until the present date, it can be said that the Exner equation is at the heart of most hydro-morphodynamic theories (Colombini, 2014).

2.2 Formation of subaqueous bedforms: State of the art review

2.2.1 Sediment transport

An important ingredient in the process of sedimentary pattern formation is sediment transport which involves the erosion and deposition of sediment particles as a result of the net effect of hydrodynamic, gravity and inter-particle forces. Depending on the ratio between the driving hydrodynamic force and the resisting gravity and bed friction forces, sediments are either stationary or in motion. There is a general consensus that the onset of particle motion and transport is controlled by the Shields number Θ. The Shields number, which is the ratio of the bottom shear stress to the apparent weight of a particle, is defined as (Shields, 1936):

$$\Theta = \frac{u_\tau^2}{\left(\rho_p/\rho_f - 1\right)|\mathrm{g}|D}, \tag{2.21}$$

where $|\mathrm{g}|$ and D are the magnitude of acceleration due to gravity and the particle diameter respectively. ρ_p/ρ_f is the particle to fluid density ratio. In a given flow, below a critical value Θ_c, the shear stress induced by the flow is unable to dislodge and entrain particles from the bed and no sediment erosion is observed. Θ_c exhibits a weak dependence on the particle Reynolds number based on the particle diameter and the friction velocity $D^+ = Du_\tau/\nu$ (Shields, 1936). A number of experimental studies on the incipient of sediment motion have shown that the value of the critical Shields number depends on many factors among which are the constituent grain packing, grain shape, polydispersity, inter-particle friction, etc.

[3] "*Morphodynmics:* a novel discipline concerned with the understanding of processes whereby sediment motion is able to shape Earth's surface" (Seminara, 2010).

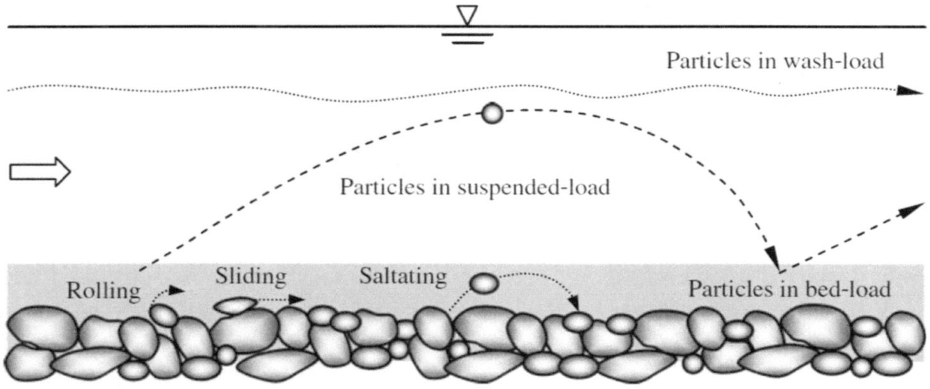

Figure 2.1: Modes of sediment transport (material copied from Dey, 2014, page 262).

In a given stream, when the non-dimensional shear stress exceeds the threshold value Θ_c, a stationary sediment bed is set in motion. Typically, larger particles are transported along the sediment bed as *bedload*. In this mode of transport, particles experience regular contact with the granular bed and move by rolling, sliding or in a series of hopping motions called saltation. On the other hand, smaller particles like fine sand generally get suspended and are transported as *suspended load*. Suspended sediment is transported by processes of advection and turbulent diffusion (García, 2008b, page 107). Figure 2.1 shows a nice schematic diagram which distinguishes the different modes of sediment transport (Dey, 2014). In the flow configurations considered in this thesis, bedload is the dominant mode of particle transport and hereafter the term 'particle flow rate' denotes mainly the rate of particle transport as bedload.

At values of the Shields number larger than Θ_c, the resulting sediment flux is usually expressed as a function of the local bed shear stress (or the excess shear stress $\Theta - \Theta_c$). A considerable amount of experimental and theoretical studies have been carried out in the past, leading to a number of (semi-) empirical predictive models for engineering purposes. For instance, there exist several algebraic expressions for the particle flux as a function of the local bed shear stress both in the laminar and turbulent flow regimes (see e.g. García, 2008a, page 72, Ouriemi et al., 2009b). Critically assessing the validity of the proposed models is a challenging task due to the complex interaction between the flow and the mobile sediment bed, and due to the dependence on multiple parameters. For similar reasons, available experimental data is widely dispersed. Moreover, controlling and determining flow parameters as well as unambiguously defining relevant quantities has been technically challenging. To the present date, experimentally accessing the complete space- and time-resolved information regarding the movement of the individual sediment grains as well as that of the flow is very challenging.

Transport of mono-dispersed, spherical particles sheared by laminar flow

In order to better understand the fundamental granular transport aspects, some studies have simplified the flow problem, at the same time retaining the underlying physics, by considering erosion of sediment bed, which is made up of mono-dispersed spherical particles, induced by shearing laminar flows (see e.g. Charru et al., 2004; Loiseleux et al., 2005; Charru et al., 2007; Ouriemi et al., 2007; Lobkovsky et al., 2008; Mouilleron et al., 2009; Ouriemi et al., 2009b; Aussillous et al., 2013). In an attempt to determine Θ_c in an unambiguous manner, Ouriemi et al. (2007) have performed an experimental investigation of the cessation of motion (which indirectly yields the threshold for the onset of motion) of spherical beads in laminar pipe flow. Systematically, they inferred that the critical Shields number has a value $\Theta_c = 0.12 \pm 0.03$, independent of the Reynolds number. This value has also been reported by other authors (Charru et al., 2004; Loiseleux et al., 2005). Note that, Θ_c is different from and larger in value than another critical Shields number which corresponds to the initiation of the motion of individual particles in an initially loosely packed granular bed. Experiments show that particles, when they are initially set in motion, move in an erratic manner by rolling over other particles, temporarily halting in troughs and then starting to move again, most of the time impacting other particles and possibly setting them in motion. After a sufficient duration, the particles rearrange and the sediment bed gets compacted. Charru et al. (2004) refer to this phenomenon as the 'armoring effect of the bed'; they explain the observed initiation of motion of particles at $\Theta \approx 0.04$ in an initially loosely packed sediment bed and describe the gradual increase of the critical shear number towards Θ_c.

At super-critical values of the Shields number, Charru and Mouilleron-Arnould (2002), applying the viscous resuspension model of Leighton and Acrivos (1986), found that the particle flux varies cubically with the Shields number. Similarly, Ouriemi et al. (2009b), considering an alternative continuum description of bedload transport and assuming a frictional rheology of the mobile granular layer, proposed an expression for the dimensionless particle flux which likewise predicts a cubic variation with the Shields number for $\Theta \gg \Theta_c$. Charru et al. (2004) on the other hand, have proposed an erosion-deposition model that predicts the saturated surface flux of sediments near the threshold (computed from the surface density of moving particles at the fluid-bed interface) to vary quadratically with Θ.

2.2.2 Initiation and evolution aspects of subaqueous patterns

Selection of the initial characteristic wavelength

Most of the previous theoretical work on the formation of ripples, dunes and similar sediment patterns is based upon the notion that a flat bed is unstable with respect to perturbations of sinusoidal shape. It was Kennedy (1963) who first studied this instability problem by considering a potential flow solution and over the years the concept has been applied by a number of researchers for a variety of flow conditions, in the laminar (Charru and Mouilleron-Arnould, 2002; Charru and Hinch, 2006a) and turbulent regime (Richards, 1980; Sumer and Bakioglu, 1984; Colombini, 2004; Colombini and Stocchino, 2011). Invoking a disparity in time scales between the flow and the bed shape modification, most of the approaches have considered the bed shape as fixed for the purpose of the analysis. The hydrodynamic problem is then coupled to the morphodynamic problem by an expression for the particle flux as a function of the local bed shear stress at a given transversal section of the flow. Linear stability analysis is then performed to determine the most amplified bedform wavelength.

It is agreed in most of the theoretical approaches that the mechanism which destabilizes a perturbed flat bed, is the phase-lag between the perturbation and the bottom shear stress which is a consequence of the fluid inertia. A balance between this destabilizing mechanism and other stabilizing effects such as gravity (Fredsøe, 1974) or phase-lag between bottom shear stress and the particle flow rate (Charru, 2006) is believed to result in instability of the bed at a certain preferred wavelength. Linear stability analysis is then carried out to predict the most unstable wavelength. Compared to experimental observations, however, predictions from most of these approaches can be broadly described as unsatisfactory, sometimes predicting pattern wavelengths which are off by an order of magnitude (Raudkivi, 1997; Langlois and Valance, 2007; Coleman and Nikora, 2009; Ouriemi et al., 2009a).

In some recent approaches, it is argued that the particle flux does not immediately adapt to a change in the local bed shear stress. Consequently, a relaxation effect is taken into account through a separate differential equation for the particle number density (Charru, 2006; Charru and Hinch, 2006a) or for the particle flux (Valance, 2005; Valance and Langlois, 2005; Fourrière et al., 2010). In these approaches, a characteristic relaxation length scale, namely the saturation length, is introduced which represents the scaling of the initial unstable bedform wavelength. The saturation length encompasses a wide spectrum of length scales as a consequence of simultaneously occurring particle transport modes (rolling, sliding, saltation, etc.) and dynamical mechanisms (erosion, rebounding, colliding, turbulence action, etc.). In modeling the saturation length, however, bed pattern initiation is commonly assumed to be dominantly governed by the slowest mechanism (thus by the largest length scale) for a given regime. For instance, for massive particles, the inertial effect might dominate and the saturation length is said to scale with the drag length l_d which is often based on Stokes drag, i.e. $l_d \approx (\rho_p/\rho_f)D$. Other saturation length models account for different hydrodynamic and granular bed interactions and introduce other length scales, such as the deposition length which represents the distance traveled by particles (moving at a certain characteristic velocity), during their deposition time $t_d \sim D/u_s$, where u_s is the particle settling velocity (Charru, 2006; Franklin and Charru, 2011). In the context of aeolian sand transport, the saltation length, which represents the distance covered by saltating grains during their flight time, is suggested to be the most important length scale (see e.g. Sauermann et al., 2001). Although approaches which include the relaxation effect into the linear stability analysis have been partially successful, they are still unable to accurately determine the initial wavelength (Langlois and Valance, 2007; Ouriemi et al., 2009a). This can probably be attributed to the lack of a fully appropriate model for the relaxation length which stems from the incomplete description of the complex particle dynamics near and inside the bed.

It is difficult to experimentally track the very instant of the bed instability. Most available experimental studies report wavelengths of the developed bedforms which have undergone a coarsening process (the temporal evolution of the initial patterns to their 'steady-state' form) or possibly have coalesced with other bedforms. There are several experimental studies (see e.g. Coleman and Melville, 1994; Betat et al., 2002; Coleman et al., 2003; Langlois and Valance, 2007; Ouriemi et al., 2009a) which report on the initial wavelength and its development. However, reported data is widely dispersed. Furthermore, it is difficult to experimentally capture the three-dimensional nature of the individual particle and fluid motion within the bed layer. As a consequence, it has been difficult in the past to experimentally assess the validity of the theoretical predictions or the various proposed phenomenological parameters.

Control parameters of bedform instability

In the literature, there seems to be no consensus concerning the parameters which control the instability of an erodible bed. Different theoretical contributions predict different thresholds for the instability based on different claimed governing parameters. In the following, we will review several relevant studies.

Richards (1980) predicts two modes of bed instability: one mode (associated with the small scale ripples) is said to be controlled by the bed roughness height, and the other mode (associated with the large scale dunes) by the flow depth. Extending the theory of Richards (1980), Sumer and Bakioglu (1984) claim that the formation of the small scale ripples is controlled by the Reynolds number based on particle Reynolds number D^+. Above a certain critical value of D^+, an erodible flat bed is said to be unstable at a wavelength which scales with the viscous length scale. Colombini and Stocchino (2011) on the other hand, following the previous analysis of Colombini (2004), argue that the formation of both ripples and dunes is controlled by the particle Reynolds number and the Shields number. The formation of dunes as a primary instability has been recently questioned by Fourrière et al. (2010) which argue that dunes are a result of nonlinear pattern coarsening. From another standpoint, the erosion-deposition model of Charru and Hinch (2006a) predicts a long-wave bed instability which is controlled by the Galileo number $Ga$and the Shields number Θ. The Galileo number is defined as

$$Ga = \frac{\sqrt{(\rho_p/\rho_f - 1)|\mathbf{g}|D^3}}{\nu} \tag{2.22}$$

and gives a measure of the ratio between gravity and viscous forces. Below a critical value of Ga, the bed is predicted to be always stable; at higher values of Ga, the bed is said to be unstable, if in addition the Shields number exceeds a certain critical value. This prediction is in contrast to the observations made in the experimental study by Ouriemi et al. (2009a), who carried out pattern evolution experiments of a flat bed in a pipe flow configuration. The authors observed that, once the bed is set in motion (threshold determined by the Shields number), the destabilization of the flat bed is solely controlled by the pipe Reynolds number. Above a critical value of the pipe Reynolds number, the flat bed destabilizes and evolves into different bedform regimes depending whether the flow is laminar or turbulent.

The multitude of proposed governing parameters is mainly a consequence of the adoption of various (semi-) empirical expressions of the particle flux as a function of the local bed shear stress. Thus, the adequacy of the various theoretical approaches lies in the ability of the different algebraic expressions to accurately describe the bed-load flux. To the present date however, the complexity of the dynamical processes involved has led to an unsatisfactory description of the bed-load flux by most available models, and thus to a high degree of uncertainty in determining which parameters control the threshold for flat bed instability.

Processes of bedform evolution

The subsequent bedform evolution is an even more complicated issue which includes aspects like coarsening (bedform wavelength and amplitude growth), coalescence (fusion of bedforms), 3D patterning (loss of bedform translational invariance). Even in simplified and controlled experiments where bedforms are essentially two-dimensional (see e.g. Betat et al., 2002; Ouriemi et al., 2009a), the evolution process is result of nonlinearly interacting complex mechanisms which are far from our understanding. Certainly, linear stability theories are not adequate in describing the bedform transient processes such as the evolution of the amplitude or the evolution of bedform morphology towards the asymmetric shape (which is even observed in the absence of flow separation). Weakly nonlinear stability theories have been proposed in some publications (see e.g. Colombini and Stocchino, 2008), but still are unable to predict or describe bedforms such as the "vortex dunes" observed in the experiments of Ouriemi et al. (2009a). The vortex dunes are characterized by flow separation and recirculation at their fronts and their initial evolution dynamics differ from those of the small scale bedforms without flow separation (Ouriemi et al., 2009a).

2.2.3 Flow over subaqueous bedforms

The flow field evolving over subaqueous patterns is statistically unsteady and two-dimensional which exhibits complex interaction with the bedforms. Best (2005) summarizes the distinct flow structures in a turbulent flow over asymmetrical bedforms as follows (cf. figure 2.2). In the region over the stoss side of the bedforms (upstream of the dune crest), fluid particles exhibit a mean acceleration and attain maximum streamwise velocity as they reach the region over the crest. This is a consequence of an increase in the favorable pressure gradient. Attached to the fluid-bed interface, an internal boundary layer develops growing over the entire

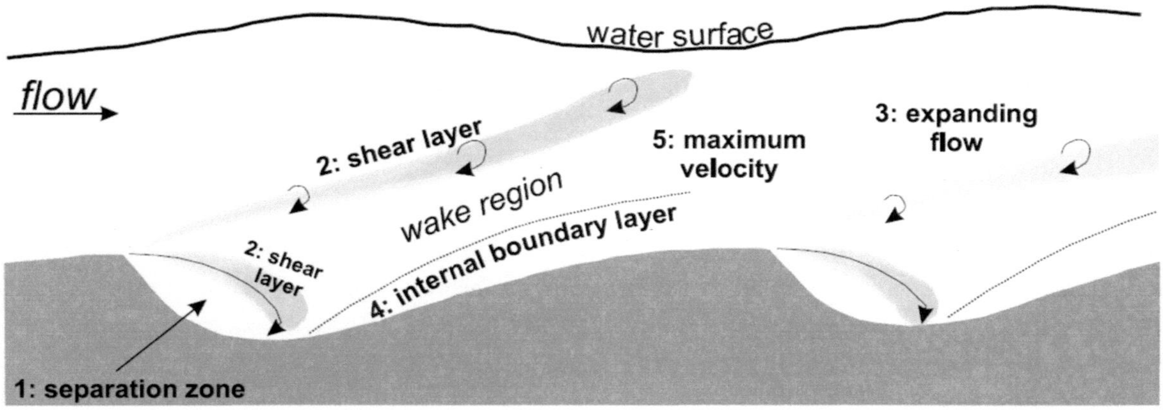

Figure 2.2: Schematic diagram of the flow over asymmetrical dunes (material copied from Best, 2005).

stoss side. On the other hand, in the flow region over the lee side of the dune (downstream of the crest), fluid particles decelerate due to an adverse pressure gradient. Generally, flow separation occurs at a point along the dune surface immediately downstream of the crest. As a result of the flow separation, a shear layer region develops which separates the bulk flow from the re-circulation region where flow reversal occurs. The re-circulation region extends several dune heights (4–6 times) downstream of the crest. The shear layer is the region where large-scale energetic turbulent structures are generated as a result of Kelvin-Helmholtz type instabilities and is characterized by high turbulent kinetic energy production region. Further downstream, the shear layer region expands and creates a wake region.

The accurate prediction of the the properties of the flow over bedforms is very crucial since the mutual interaction between the flow and the bedforms defines important aspects of sedimentary patterns such as the spatial variation of the bottom boundary shear stress along the bedforms and its relation to the sediment transport rate (Best, 2005). However, for obvious reasons, predicting such a complex flow over space and time varying bedforms is very challenging. A number of experimental and numerical studies have simplified the problem by considering flow over fixed dunes (e.g. McLean and Smith, 1986; Nelson and Smith, 1989; Nelson et al., 1993; McLean et al., 1994; Bennett and Best, 1995; Stoesser et al., 2008; Omidyeganeh and Piomelli, 2011). Although such studies have been useful in elucidating the key features of the flow, they fail to provide information on the 'four-way' coupled interaction between the flow and the individual sediment particles of the evolving sediment bed. There are very few studies which have attempted to characterize the flow over an evolving bed (see e.g. Charru and Franklin, 2012; Khosronejad and Sotiropoulos, 2014).

2.2.4 Knowledge gaps

The above brief literature review has shown that the physical mechanisms involved in the process of sedimentary pattern formation are not yet fully understood. We believe that addressing the following identified knowledge gaps will contribute to the elucidation of these mechanisms.

- A complete four-dimensional (space and time) information about the driving turbulent flow field, for instance the distribution of the basal shear stress which is responsible for the destabilization of a flat erodible bed, is not available.

- Detailed studies which correlate the individual or collective sediment grain motion to the associated turbulent flow and to the evolving sediment bed, are not available. This has resulted in an unsatisfactory description of models such as for the sediment flux, and has limited the understanding of the mechanisms involved in the bedform evolution processes.
- The accuracy of determining the preferred wavelength during the initial bed instability, either theoretically or experimentally, has been unsatisfactory.
- Refined studies which are necessary in evaluating proposed models or phase diagrams such as that of Ouriemi et al. (2009a) are not sufficiently available.

Chapter 3

Computational approach

3.1 Numerical method

3.1.1 Direct numerical simulation

In computational fluid dynamics, turbulent flows are treated by performing one of the following simulation strategies. Direct numerical simulation (DNS), large eddy simulation (LES) or Reynolds averaged Navier-Stokes equations (RANS) modeling. In DNS, the Navier–Stokes equations are solved directly without a turbulence model, on a computational grid which is resolved enough to capture the smallest scales of the flow, and at the same time large enough to capture the large scales. Moreover, time step of integration also has to small enough to capture the smallest time scales. As a result, performing DNS is computationally very expensive. In LES, the Navier–Stokes equations are first derived for a filtered velocity field in which the small scales are smoothed out. The filtered equation are then solved on a grid which resolves the large scales of the flow containing most of the energy spectrum, while the unresolved small scales are modeled. Generally, performing LES is computationally efficient than performing DNS. On the other hand, as the name implies, RANS models solve the Reynolds averaged Navier–Stokes equations in which the Reynolds stress tensor is modeled by providing a closure term. The above description of the different approaches is very incomplete. A detailed overview can be found for example in the textbook by Pope (2000). In the present thesis, the equation of fluid motion are numerically solved by performing DNS.

3.1.2 Immersed boundary method

The most faithful methods of simulating particulate flow problems are those in which the interface between the phases is precisely defined by employing a curvilinear *body-fitted* grid around the immersed bodies. The Navier-Stokes equations are then solved in a time-dependent domain by applying the no-slip condition at the interfaces (Mittal and Iaccarino, 2005). In such approach, determining the interaction between phases is straightforward and can be accomplished with a very high accuracy. However, body-fitted approaches are not practically feasible for simulating particulate flow problems specially when multiple moving suspended particles are involved. This is mainly due to the enormous computational cost of grid regeneration at each time step to define the new fluid domain (Prosperetti and Tryggvason, 2007).

The immersed boundary method (IBM) is a different approach to the treatment of fluid-solid interaction which avoids the drawbacks of the body-fitted methods. The basic idea of IBM is to solve the Navier-Stokes equations on a fixed Cartesian grid spanning throughout the entire domain Ω comprising the fluid domain Ω_f and the domain occupied by the suspended particles Ω_s. The presence of the particles is accounted for by adding appropriate source terms to the Navier-Stokes equations which serves to impose the no-slip condition at the fluid-solid interface. Peskin was first to introduce the idea of the IBM in 1972 to study blood flows inside the heart (Peskin, 2002), and since then the method has been subject to a lot of development and a number of variants of the method exist today (Mittal and Iaccarino, 2005). The main aspect of the different variants is the way the forcing term is formulated.

The numerical method employed in the present thesis is a formulation of the immersed boundary method developed by Uhlmann (2005a). The variant proposed by Uhlmann (2005a) is suited for the treatment of rigid solid immersed objects by employing a direct forcing approach, where a localized volume force term is added to the momentum equations. The additional forcing term is explicitly computed at each time step as a function of the desired particle positions and velocities, without recurring to a feed-back procedure as in the original IBM. The interpolation of variables from Eulerian grid nodes to the particle-related Lagrangian positions and the spreading of the computed force terms back to the Eulerian grid is carried out by employing the regularized delta functions introduced by Peskin (2002). This procedure yields a smooth temporal variation of the hydrodynamic forces acting on individual particles while these are in arbitrary motion with respect to the fixed grid. To improve the efficiency of the code, forcing is applied only to the surface of the spheres, leaving the flow to freely develop in the space inside the particles. Uhlmann (2005a) have demonstrated that this procedure has negligible effect on particle motion. Since particles are free to visit any point in the computational domain and in order to ensure that the regularized delta function verifies important identities (such as the conservation of the total force and torque during interpolation and spreading), a Cartesian grid with uniform isotropic mesh width is employed.

The modified Navier-Stokes equations are discretized based on the standard fractional step method for the incompressible Navier-Stokes equations. The temporal discretization is semi-implicit, based on the Crank-Nicolson scheme for the viscous terms and a low-storage three-step Runge-Kutta procedure for the non-linear part (Verzicco and Orlandi, 1996). The spatial operators are evaluated by central finite-differences on a staggered grid. The temporal and spatial accuracy of this scheme are of second order.

The DNS code has been validated on a whole range of benchmark problems (Uhlmann, 2004; Uhlmann, 2005a; Uhlmann, 2005b, 2006; Uhlmann and Dušek, 2014), and has been previously employed for the simulation of several flow configurations (Uhlmann, 2008; Chan-Braun et al., 2011; García-Villalba et al., 2012; Kidanemariam et al., 2013; Chan-Braun et al., 2013).

3.1.3 Inter-particle collision model

A common approach of numerically treating dense systems such as dry/wet granular flows is to represent the system by a large number of discrete elements or particles. The method is called discrete element modeling (DEM). Each of the constituent particles interact with one another by short-range collision forces. The basic idea of the approach is to model these interaction forces and numerically solve the trajectory of each element. Various collision models have been proposed which are usually based either on the hard-sphere or the soft-sphere approaches (Crowe and Sommerfeld, 1998).

The hard-sphere model is common in molecular dynamics. In the model, particles are assumed to be rigid and exchange momentum during instantaneous binary collision events (see e.g. Tanaka and Tsuji, 1991; Hoomans et al., 1996; Foerster et al., 1994). The details of the contact, which happen in finite duration of time, are not resolved, rather the integral form of the equation of rigid body motion (impulse-momentum equation) are solved during a certain time step (assuming particles translate and rotate linearly within one time step). If, within a given time step, a collision between two particles occurs, the time and their position at contact is determined. Based on the intermediate position and post-collision velocities, particle positions at the end of the time step are updated. The coefficient of friction (which according to Coulomb's friction law is the ratio of friction force to normal force exerted between two bodies) and the coefficient of restitution (which is defined as the ratio of particle normal velocities before and after a collision event) are used to relate the pre- and post-collision transational and angular velocities. The simple formulation of the model and the independence of the global computational time step on the contact details are the advantages of this model (Crowe and Sommerfeld, 1998). Moreover, the friction and restitution coefficients of a given material can be experimentally determined. However, the main drawback of the model is the fact that it is unable to resolve details on the collision dynamics.

On the other hand, the soft-sphere model is capable of treating contact dominated dense systems where particles may stay in (multiple) contact for a finite duration of time. The model solves the differential form of the Newton equations of motion and thus resolves the particle motion even during the contact period (see e.g. Cundall and Strack, 1979; Tsuji et al., 1993; Wachs, 2009). In the soft-sphere context, the physical infinitesimal deformation of particles during contact is indirectly considered by admitting an overlap region between the contacting particles, where this admitted overlap thickness is very small compared to the dimension of the particles. The contact forces, which are assumed to be functions of the overlap thickness and/or the relative velocities between contacting particles, are computed based on mechanical models such as springs, dashpots and sliders. These mechanical models are intended to mimic the deformation, damping and frictional effects between particles. The main drawback of the soft-sphere approach is the computational burden due to the time step which should in principle be small enough to resolve the duration of the contact (Cundall and Strack, 1979; Crowe and Sommerfeld, 1998). Moreover, the model is subject to numerical stiffness and damping coefficients, which need to be calibrated from empirical and analytical expressions which take into account material properties of the contituent particles such as the restitution coefficient, modulus of elasticity etc.

Particle-particle or particle-wall encounters have been previously treated in the DNS code by implementing the artificial repulsion force model of Glowinski et al. (1999) which is commonly used in the simulations of dilute particulate flows where particle collisions are relatively infrequent. The main purpose of the above mentioned model is to prevent non-physical overlap of particles while frictional loses during the inter-particle or wall-particle contact are not accounted for. In dense systems such as sediment transport in laminar/turbulent flows however, inter-particle contact, in addition to the fluid-particle interaction, are expected to be significant. For this purpose, we employ a discrete element model which mimics the basic collision dynamics between the immersed spherical particles instead of an artificial repulsion model such as Glowinski et al. (1999). The DEM is developed by adopting a standard soft-sphere model which is based on a linear mass-spring-damper system to model the collision forces (cf. appendix A). More specifically, let us consider the ith particle (of a system with a total number of N_p particles) which has radius $R^{(i)}$ and which, at time t, its center is located at position $\mathbf{x}_p^{(i)}$. The total collision force acting on the particle as a result of its contact with neighboring particles is defined as:

$$\mathbf{F}_c^{(i)}(t) = \sum_{j=1}^{N_p} \left(\mathbf{F}_{el}^{(i,j)}(t) + \mathbf{F}_d^{(i,j)}(t) + \mathbf{F}_t^{(i,j)}(t)\right) I_c^{(i,j)}(t) \,. \tag{3.1}$$

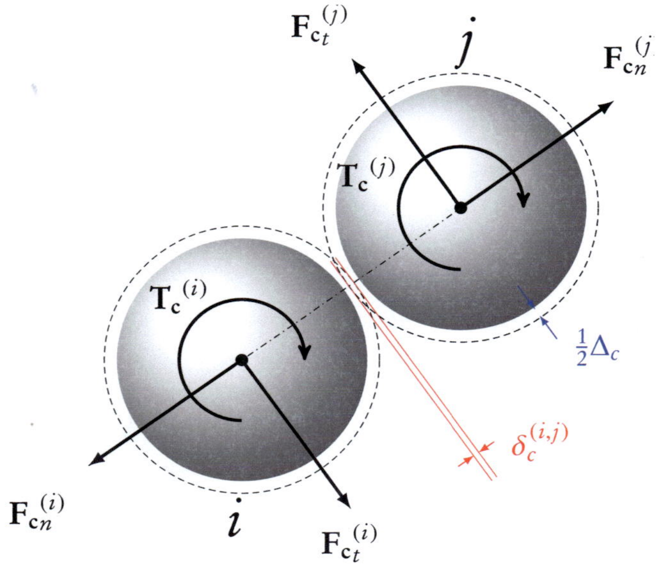

Figure 3.1: Schematic showing the collision force $F_c^{(i)}$ and torque $T_c^{(i)}$ acting on particle with index i during contact with particle j. The subscripts n and t indicate the normal and tangential components, respectively; Δ_c denotes the force range, and $\delta_c^{(i,j)}$ is the overlap length.

$I_c^{(i,j)}$ is a function which indicates whether a particle pair with indices i,j (where $i \neq j$) is in contact, viz.

$$I_c^{(i,j)}(t) = \begin{cases} 1 & \text{if} \quad |\mathbf{x}_p^{(j)}(t) - \mathbf{x}_p^{(i)}(t)| \leq R^{(i)} + R^{(j)} + \Delta_c \,, \\ 0 & \text{else} \,. \end{cases} \tag{3.2}$$

In the relation (3.2), Δ_c denotes an admitted gap of the order of the grid size Δx between the particle surfaces to account for the distance over which the interface is "smeared out" in the discrete system (i.e. related to the width of the support of the regularized delta function employed in the immersed boundary method). The three individual force contributions entering the total contact force in (3.1) are modeled as follows. The elastic part of the normal force is a linear function of the overlap between the particles. Its value (acting on the ith particle) is given by:

$$\mathbf{F}_{el}^{(i,j)}(t) = -k_n\, \delta_c^{(i,j)}(t)\, \mathbf{e}_n^{(i,j)}(t) \,, \tag{3.3}$$

where k_n is a stiffness coefficient, δ_{ij} is the penetration (ovelap) length defined as

$$\delta_c^{(i,j)}(t) = R^{(i)} + R^{(j)} + \Delta_c - |\mathbf{x}_p^{(j)}(t) - \mathbf{x}_p^{(i)}(t)| \,, \tag{3.4}$$

and $\mathbf{e}_n^{(i,j)}$ is the unit normal vector along the line connecting the particles centers pointing from $\mathbf{x}_p^{(i)}$ to $\mathbf{x}_p^{(j)}$:

$$\mathbf{e}_n^{(i,j)}(t) = \frac{\mathbf{x}_p^{(j)}(t) - \mathbf{x}_p^{(i)}(t)}{|\mathbf{x}_p^{(j)}(t) - \mathbf{x}_p^{(i)}(t)|} \,. \tag{3.5}$$

The dissipative part of the normal force acting on the ith particle is defined as:

$$\mathbf{F}_d^{(i,j)}(t) = -c_{dn}\, \mathbf{u}_{r,n}^{(i,j)}(t) \,, \tag{3.6}$$

where c_{dn} is a normal damping coefficient and $\mathbf{u}_{r,n}^{(i,j)}$ is the normal component of the relative velocity of particle i with respect to particle j at the contacting edges (points A and B in figure 3.1):

$$\mathbf{u}_r^{(i,j)}(t) = \mathbf{u}_p^{(i)}(t) - \mathbf{u}_p^{(j)}(t) + \boldsymbol{\omega}_p^{(i)}(t) \times R^{(i)}\mathbf{e}_n^{(i,j)}(t) + \boldsymbol{\omega}_p^{(j)}(t) \times R^{(j)}\mathbf{e}_n^{(i,j)}(t)\,, \tag{3.7}$$

$$\mathbf{u}_{r,n}^{(i,j)}(t) = \left(\mathbf{e}_n^{(i,j)}(t) \cdot \mathbf{u}_r^{(i,j)}(t)\right)\mathbf{e}_n^{(i,j)}(t). \tag{3.8}$$

The force acting tangentially (on the ith particle) applied at the contact points between particles i and j, is computed according to the following expression:

$$\mathbf{F}_t^{(i,j)}(t) = \begin{cases} -\left[\min\{\mu_c|\mathbf{F}_{el}^{(i,j)}(t) + \mathbf{F}_d^{(i,j)}(t)|, c_{dt}|\mathbf{u}_{r,t}^{(i,j)}(t)|\}\right]\mathbf{e}_t^{(i,j)}(t) & \text{if} \quad |\mathbf{u}_{r,t}^{(i,j)}(t)| \neq 0\,, \\ 0 & \text{else}\,. \end{cases} \tag{3.9}$$

Relation (3.9) expresses the fact that the tangential frictional force with damping coefficient c_{dt} is proportional to the tangential component of the relative velocity at the contact point (cf. below); it is, however, limited by the Coulomb friction (i.e. the condition of traction) which is proportional to the normal force acting at the same contact point with a friction coefficient μ_c. The tangential component of the relative velocity $\mathbf{u}_{r,t}^{(i,j)}(t)$ is given by the following relation:

$$\mathbf{u}_{r,t}^{(i,j)}(t) = \mathbf{u}_r^{(i,j)}(t) - \mathbf{u}_{r,n}^{(i,j)}(t)\,, \tag{3.10}$$

while the tangential unit vector $\mathbf{e}_t^{(i,j)}$ is defined as:

$$\mathbf{e}_t^{(i,j)}(t) = \frac{\mathbf{u}_{r,t}^{(i,j)}(t)}{|\mathbf{u}_{r,t}^{(i,j)}(t)|}, \quad \forall \quad |\mathbf{u}_{r,t}^{(i,j)}(t)| \neq 0\,. \tag{3.11}$$

The tangential component of the total collision force generates a torque $\mathbf{t}_c^{(i)}$ which acts on particle i, viz.

$$\mathbf{t}_c^{(i)}(t) = R^{(i)}\mathbf{e}_n^{i,j} \times \sum_{j=1}^{N_p} \mathbf{f}_c^{(i,j)}(t) I_c^{(i,j)}(t)\,. \tag{3.12}$$

The model described in (3.1-3.12) introduces four parameters affecting the collision process namely: k_n, c_{dn}, c_{dt} and μ_c. From an analytical solution of the linear mass-spring-damper system in an idealized configuration (binary collision of uniformly translating spheres in vacuum and in the absence of external forces), a relation between the normal stiffness coefficient k_n and the normal damping coefficient c_{dn} can be formed. For this purpose a dry restitution coefficient is defined ε_d as the ratio between the post- and pre-collision normal relative velocities in vacuum as

$$\varepsilon_d = -\frac{|\mathbf{u}_{r,n}^{(i,j)}|_{post}}{|\mathbf{u}_{r,n}^{(i,j)}|_{pre}}\,. \tag{3.13}$$

It can then be shown that the following relation holds between the normal damping and stiffness coefficient:

$$c_{dn} = -2\sqrt{M_{ij}k_n}\frac{\ln \varepsilon_d}{\sqrt{\pi^2 + \ln^2 \varepsilon_d}}\,, \tag{3.14}$$

where $M^{(l)}$ is the mass of the lth particle and $M_{ij} = M^{(i)}M^{(j)}/(M^{(i)} + M^{(j)})$ is the reduced mass of particle pair i,j (see e.g. Crowe and Sommerfeld, 1998).

A crucial parameter of the collision model described above is the duration of a collision event because it determines the time step required to numerically integrate the Newton equations of motion such that the collision dynamics is resolved. For instance, in the context of dry granular flow, Cleary and Prakash (2004) state that 15 to 25 time steps per collision duration are required for accurate integration of particle motion as a result of collision forces. As a reference value, the collision duration for the ideal configuration mentioned above is analytically given by (cf. appendix A):

$$T_c = \frac{2\pi M_{ij}}{\sqrt{4M_{ij}k_n - c_{dn}^2}}. \tag{3.15}$$

In all the simulations of the present study, we have adopted a conservatively small collision time step $\Delta t_{sub} = 10^{-5}$ which yields $196 \lesssim T_c/\Delta t_{sub} \lesssim 878$ for the particle-wall collision tests (cf. section 3.2) and $T_c/\Delta t_{sub} \approx [330; 950]$ for the shear induced sediment erosion simulations (cf. chapter 4).

3.1.4 Time discretization of the equations of particle motion

In the context of the soft-sphere approach, as was mentioned above, the DEM solver requires very small time steps to resolve the contact forces during particle-particle or particle-wall collisions due to the large value of stiffness coefficient typically chosen. This is much smaller than the typical time step of the fluid solver. In order to avoid this restriction, one approach is to apply a subs-stepping strategy by which the Newton equations for particle motion are solved with a smaller time step than that used for solving the Navier-Stokes equations (see e.g. Wachs, 2009). Recently Kempe and Fröhlich (2012) have proposed a strategy of artificial collision time stretching, matching the time scale of the collision to that of the fluid. In the present work, we have adopted the former strategy of separately resolving the two time scales. Sub-stepping loosely amounts to freezing of the flow field for the duration of a certain number of time steps N_{sub} (where $N_{sub} = \Delta t/\Delta t_{sub}$, with Δt and Δt_{sub} the time steps of the fluid and the particle phase, respectively), between two consecutive updates of the fluid phase (cf. appendix B for details). It should be noted that the particle related computational load is not critically limited by the collision treatment in our computational implementation. This allowed us to adopt a conservatively small value of Δt_{sub} which should be sufficient even for generic collisions happening in configurations beyond the simple case of an idealized head-on collision leading to equation (3.15). In the cases treated in the present work, the number of sub-steps adopted ranges from a minimum value $N_{sub} = 105$ up to a maximum value of $N_{sub} = 240$.

3.2 Collision of a sphere with a wall in a viscous fluid

Recently, several studies have emerged which propose a more realistic modelling of the inter-particle collisions, at the same time fully resolving the background flow even in the near vicinity of individual grains (see e.g. Yang and Hunt, 2008; Wachs, 2009; Li et al., 2011; Simeonov and Calantoni, 2012; Kempe and Fröhlich, 2012; Brändle de Motta et al., 2013). For the purpose of validation of the DEM-DNS coupling, most of these studies have considered a benchmark test case where a single finite-size spherical particle is allowed to collide with a wall or with another particle in a viscous fluid. Experimental studies have shown that when a particle freely approaches and collides with another particle or a wall, in addition to energy dissipation from the solid-solid contact, it loses energy as a result of the work done to squeeze out the fluid from the gap between the contacting edges, thereby decelerating it prior to the contact (Joseph et al., 2001; Gondret et al., 2002; Cate et al., 2002; Joseph and Hunt, 2004; Yang and Hunt, 2006). Similarly, loses occur during its rebounding phase.

The effect of the viscous fluid on the bouncing behavior of the particle is classically characterized by an effective coefficient of restitution ε which is the ratio of particle's pre- and post-collision velocities aligned normal to the wall. Thus ε accounts for the total energy dissipation both from viscous fluid resistance as well as from the actual contact with the wall in contrast to the dry coefficient of restitution defined in the same way but for collisions happening in vacuum. It is well established that the Stokes number, defined as

$$St = \frac{1}{9}(\rho_p/\rho_f)Re_p \tag{3.16}$$

where Re_p is the particles Reynolds number based on its diameter and it velocity before impact, is the relevant parameter which determines the degree of viscous influence on the bouncing behavior of a particle. At large values of the Stokes number (above $St = 1000$ say), the effect of the fluid on the bouncing becomes negligible and ε approaches the dry coefficient of restitution. On the other hand, at small values of the Stokes number ($St \lesssim 10$), viscous damping is so large that no rebound of the particle is observed. The availability of Various sets of experimental data with respect to the variation of ε as a function of the Stokes number (see e.g. Joseph et al., 2001; Gondret et al., 2002) as well as the simplicity of the setup, is most probably the main reason that has lately led this classical configuration to serve as a benchmark for the validation of numerical approaches which model the collision dynamics of finite-size objects fully immersed in a viscous fluid.

3.2.1 Computational setup and parameter values

For validation purposes we have simulated the case of an isolated sphere settling on a straight vertical path under the action of gravity before colliding with a plane horizontal wall. The chosen setup is similar to the experiment of Gondret et al. (2002). Our computational domain has periodic boundary conditions in the wall-parallel directions x and z, and a no-slip condition is imposed at the domain boundaries in the wall-normal direction y. Gravity is directed in the negative y direction. A spherical particle of diameter D is released from a position near the upper wall located at a wall-normal distance L_y from the bottom wall. From dimensional considerations this problem is characterized by two dimensionless numbers, e.g. the solid-to-fluid density ratio ρ_p/ρ_f, and the Galileo number Ga. However, the data for the particle motion under normal collision with a wall is well described as a function of the Stokes number St alone (Joseph et al., 2001; Gondret et al., 2002). The Stokes number can be defined from the terminal settling velocity v_{pT} of the sphere.

In the present work we have simulated 24 different combinations of values for the density ratio and the Galileo number. These values are given in table 3.1 along with the corresponding values of the terminal Reynolds number and of the Stokes number. The present simulations cover the range of $St = 4.7 - 1880$. The size of the computational domain as well as the grid resolution (expressed as the ratio between the particle diameter and the mesh width, $D/\Delta x$) are given in table 3.2. It should be pointed out that, although the flow remains axisymmetric at all times, the simulations were done with a full three-dimensional Cartesian grid with uniform mesh size. Three additional simulations (termed $C08^b$, $C08^c$, $C08^d$) with the same physical parameter values as in case C08 have been run in order to check the sensitivity of the results with respect to the spatial resolution and the choice of the value for the force range Δ_c (cf. table 3.2). Concerning the parameters for the solid-solid contact model, the following values were chosen. The stiffness parameter k_n, normalized by the submerged weight of the particle divided by its diameter, varies in the range of $2 \cdot 10^4 - 2 \cdot 10^5$, cf.

Case	ρ_p/ρ_f	Ga	Re_p	St	k_n^*	domain
C01	2.0	30.9	21.2	4.7	40053	D1
C02	2.5	37.8	28.4	7.9	33377	D1
C03	3.0	43.7	34.9	11.6	30039	D1
C04	3.5	48.8	40.9	15.9	28037	D1
C05	4.0	53.5	46.4	20.6	26702	D1
C06	5.0	61.8	56.6	31.4	25033	D1
C07	6.0	69.1	66.0	44.0	24032	D2
C08	7.0	75.6	74.7	58.1	23364	D2
C09	8.0	81.7	82.9	73.7	22887	D2
C10	9.0	87.3	90.6	90.6	22530	D2
C11	4.0	37.8	28.4	12.6	26702	D2
C12	5.0	43.7	34.9	19.4	25033	D2
C13	6.0	48.8	40.8	27.2	23969	D2
C14	8.0	57.8	51.6	45.9	22843	D2
C15	14.0	78.7	78.5	122.2	215427	D3
C16	16.0	84.6	86.4	153.6	213406	D3
C17	18.0	90.0	93.9	187.7	211675	D3
C18	20.0	95.2	101.0	224.5	210474	D3
C19	30.0	75.1	73.9	246.3	206629	D3
C20	40.0	87.1	90.2	400.7	204997	D3
C21	50.0	97.7	105.0	583.3	204031	D3
C22	70.0	115.9	132.0	1026.7	202939	D3
C23	90.0	131.6	157.3	1573.2	202162	D3
C24	100.0	138.8	169.1	1878.3	201970	D3
C08^b	7.0	75.6	74.7	58.1	23364	D2^b
C08^c	7.0	75.6	75.4	58.6	23364	D2^c
C08^d	7.0	75.6	75.4	58.6	23364	D2^d

Table 3.1: Physical parameters of the different simulations corresponding to the vertically-oriented normal collision of a single particle with a wall in a viscous fluid. ρ_p/ρ_f is the particle-to-fluid density ratio; Ga is the Galileo number; Re_p and St are the particle's Reynolds number and the Stokes number based on the particle's terminal velocity; the normalized stiffness constant in the normal collision force is defined as $k_n^* = k_n/((\rho_p - \rho_f)|\mathbf{g}|V_p/D)$. The computational domain size and numerical parameters corresponding to the abbreviation in column 7 are listed in table 3.2.

Domain	$[L_x \times L_y \times L_z]/D$	$N_x \times N_y \times N_z$	$D/\Delta x$	$\Delta_c/\Delta x$
D1	$6.4 \times 25.6 \times 6.4$	$128 \times 513 \times 128$	20	2
D2	$6.4 \times 51.2 \times 6.4$	$128 \times 1025 \times 128$	20	2
D2^b	$6.4 \times 51.2 \times 6.4$	$128 \times 1025 \times 128$	20	1
D2^c	$6.4 \times 51.2 \times 6.4$	$192 \times 1537 \times 192$	30	2
D2^d	$6.4 \times 51.2 \times 6.4$	$192 \times 1537 \times 192$	30	1
D3	$6.4 \times 102.4 \times 6.4$	$128 \times 2049 \times 128$	20	2

Table 3.2: Numerical parameters of the particle-wall rebound simulations. L_i and N_i are the domain length and number of grid points in the ith coordinate direction, respectively; Δx is the grid spacing and Δ_c is the force range.

table 3.1. With this choice the maximum penetration distance recorded in the different simulations measures only a few percent of the force range Δ_c (i.e. $\max(\delta_{ij}(t)) \leq 0.05\Delta_c$). In two of the cases (C08^b and C08^d the maximum penetration reaches a value of approximately $0.2\Delta_c$, which still only corresponds to one percent of the respective particle diameter. The dry coefficient of restitution was set to a value $\varepsilon_d = 0.97$ corresponding e.g. to dry collisions of steel or glass spheres on a glass wall (Gondret et al., 2002). The normal damping coefficient c_n was determined according to relation (3.14).

3.2.2 Restitution coefficient versus Stokes number

Figure 3.2*a* shows the wall-normal particle position (scaled with the particle diameter) as a function of time (normalized with the reference time D/v_{pT}). It is seen that in all simulated cases the particle has attained its terminal velocity before feeling the hydrodynamic influence of the solid wall. After rebound the maximum distance between the wall and the closest point on the particle surface varies from a negligibly small value ($0.004D$) for case C01 up to approximately $40D$ for case C24. In three cases (C01, C02, C03) the maximum rebound height was below 0.04 times the particle diameter such that the results can be qualified as 'no rebound'; the corresponding Stokes numbers are $St = 4.7, 7.9, 11.6$, respectively, which is in agreement with the general observation that the bouncing transition occurs at $St_c \approx 10$ (Gondret et al., 2002). A typical time evolution of the wall-normal particle velocity is shown in figure 3.2*b*. The figure shows the steady settling regime before impact and the subsequent rapid deceleration phase which is followed by an acceleration phase as a result of the action of the contact force. In figure 3.2*b* one can also observe the effect of the hydrodynamic forces the particle experiences just before and after the collision event.

One of the principal quantities of interest in this case is the effective coefficient of restitution ε, defined as the ratio of the particle velocity values before and after the collision, viz:

$$\varepsilon = -\frac{v_{pR}}{v_{pT}}\,. \tag{3.17}$$

In collision experiments in a viscous fluid the determination of the rebound velocity v_{pR} is a somewhat delicate issue, and no unique definition seems to exist in the literature. In laboratory experiments this quantity is typically computed from measured particle position data, involving some kind of gradient computation based on the records in the direct vicinity of the wall. Therefore, the result may be sensitive to the temporal resolution of the measuring device. In the experiment of Gondret et al. (2002) the time interval between successive images was $2\,ms$ which means the first data point after impact was recorded approximately at a time of $0.3D/v_{pT}$ after the particle loses contact with the wall (under typical conditions of their experiment). In the experiment of Joseph et al. (2001), which featured a pendulum set-up with a horizontal impact on a vertical wall, the temporal resolution was comparably higher, corresponding to approximately $0.03D/v_{pT}$. In both cases the temporal resolution is significantly larger than the duration of the actual collision. As a consequence, the measured rebound velocity v_{pR} (and therefore the effective coefficient of restitution ε) takes into account to some extent the fluid-induced damping after the particle-wall contact.

In the present work we measure the rebound velocity at a predefined time after the particle loses contact with the wall. More precisely, we define

$$v_{pR} = v_p(t_1 + t_R)\,, \tag{3.18}$$

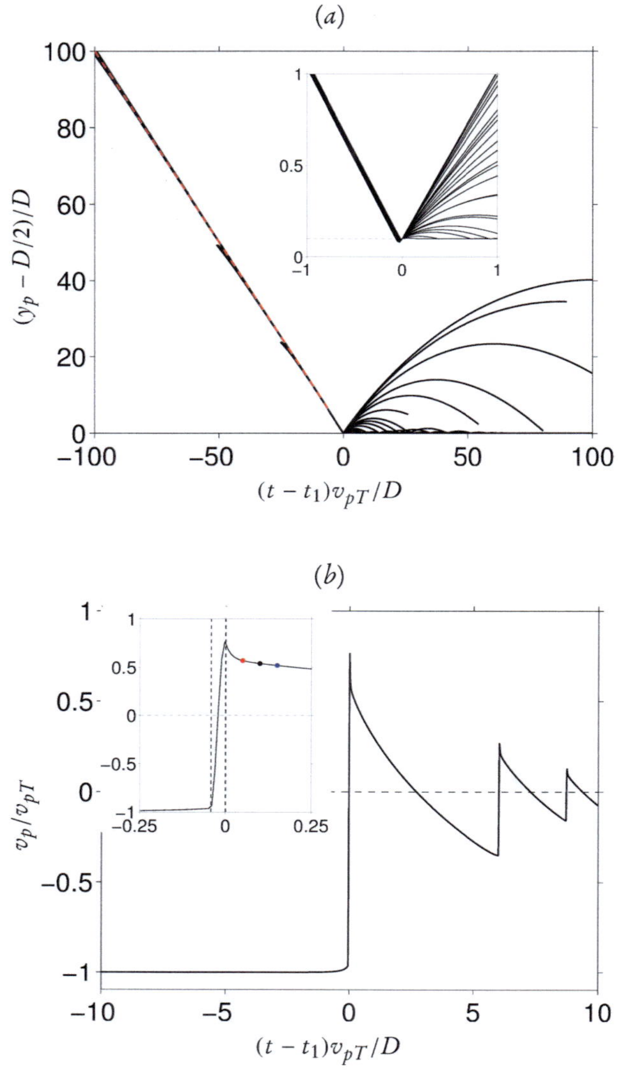

Figure 3.2: (*a*) Wall-normal position of the lowest point of the sphere $y_p - D/2$ as a function of time. In each case the time coordinate is shifted such that all particle impacts coincide with the origin of the graph. The red dotted line has a slope of -1, corresponding to motion at the terminal velocity. The horizontal dashed line in the inset indicates the Δ_c offset which is the extent of the collision force. (*b*) Wall-normal velocity of the particle in case C08 as a function of time. The symbols mark the particle rebound velocity for different values of the time delay: (▲) $t_R = 0.05D/v_{pT}$; (●) $t_R = 0.1D/v_{pT}$; (■) $t_R = 0.15D/v_{pT}$. The vertical dashed lines mark the interval during which the collision force is non-zero.

where t_1 is the instant in time when the contact force first becomes zero after the particle-wall collision, and t_R is a prescribed delay which was set to the value of $t_R = 0.1D/v_{pT}$. Note that the chosen value for the delay is comparable to the one used in the work of Gondret et al. (2002). The insert in figure 3.2*b* shows the time evolution of the particle velocity around the collision interval in the present case C08, where the rebound velocity obtained from the definition in (3.18) is marked by a symbol. Also shown in that figure are two alternative choices of delay times ($t_R\, v_{pT}/D = 0.05$ and 0.15) which will be further discussed below. The effective coefficient of restitution computed according to (3.17) is shown as a function of the Stokes number in figure 3.3. A very good match with the data points provided by the experimental measurements

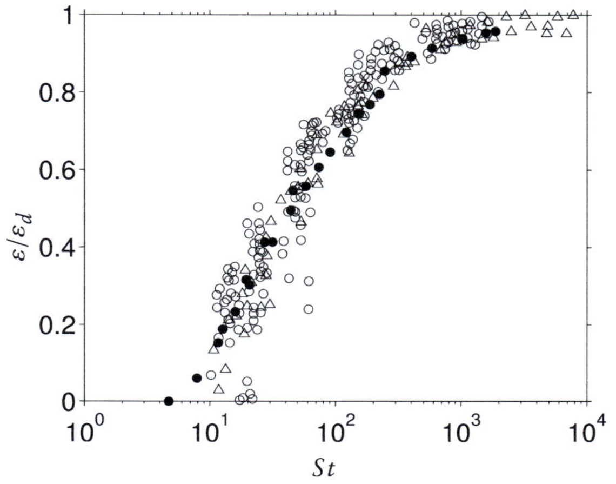

Figure 3.3: Ratio between the effective coefficient of restitution ε and the dry value ε_d as a function of the Stokes number. The present results are indicated by the symbols (●), and the rebound velocity has been computed from (3.18) with $t_R = 0.1D/v_{pT}$. Experimental data of Joseph et al. (2001) and Gondret et al. (2002) are marked by the symbols (○) and (△) respectively.

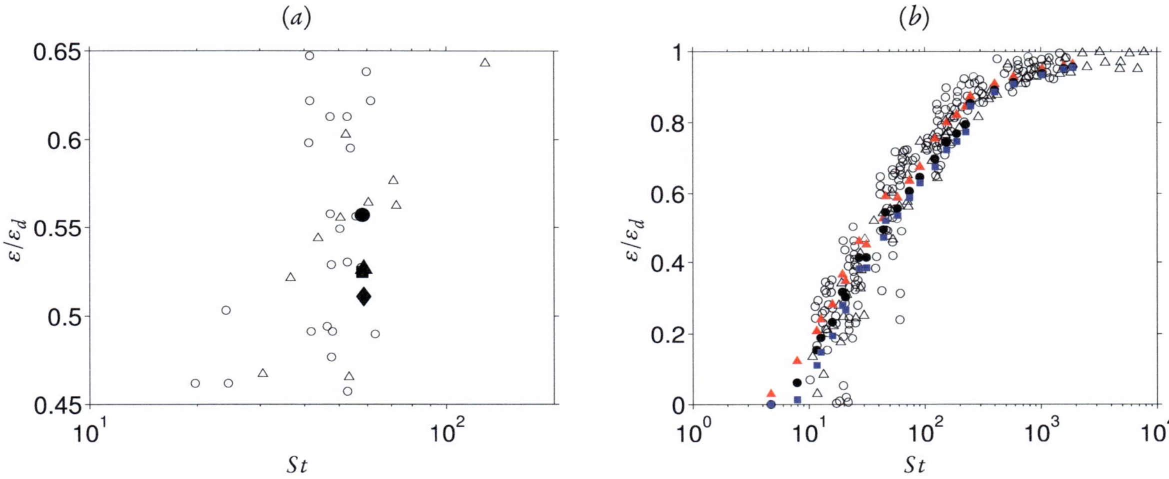

Figure 3.4: (a) Sensitivity of the computed effective coefficient of restitution with respect to the numerical parameters for $\rho_p/\rho_f = 7$ and $Ga = 75.6$: (●) case C08 ($D/\Delta x = 20$, $\Delta_c/\Delta x = 2$); (■) case C08^b ($D/\Delta x = 20$, $\Delta_c/\Delta x = 1$); (▲) case C08^c ($D/\Delta x = 30$, $\Delta_c/\Delta x = 2$); (♦) case C08^d ($D/\Delta x = 30$, $\Delta_c/\Delta x = 1$). (b) Sensitivity of ε with respect of the chosen value for the delay time t_R for all simulations: (▲) $t_R = 0.05D/v_{pT}$; (●) $t_R = 0.1D/v_{pT}$; (■) $t_R = 0.15D/v_{pT}$. In both graphs the symbols (○) and (△) correspond to the experimental data of Joseph et al. (2001) and Gondret et al. (2002), respectively.

of Gondret et al. (2002) and those of Joseph et al. (2001) can be observed over the whole parameter range, with an exponential increase in ε beyond $St_c \approx 10$. This result demonstrates that the comparatively simple collision model employed in the present work is capable of accurately reproducing the fluid-mediated impact of a spherical particle on a solid wall in the framework of an immersed boundary technique.

The results obtained in the additional simulations C08^b, C08^c, C08^d are shown in figure 3.4(a) in comparison to the original case C08 (all four simulations are for $\rho_p/\rho_f = 7$ and $Ga = 75.6$). In simulation C08^b the grid resolution is kept the same as in case C08 ($D/\Delta x = 20$), but the force range Δ_c is divided by two. The result is a reduction of the effective coefficient of restitution by approximately 6%. Nearly the same effect is obtained in case $C08^c$ where the force range is maintained constant (in multiples of the mesh width), but the spatial resolution is increased to $D/\Delta x = 30$. When setting the smaller value for the force range ($\Delta_c/\Delta x = 1$) and simultaneously choosing the decreased mesh width ($D/\Delta x = 30$) as in case $C08^d$, a reduction of the value of ε by approximately 9% is obtained. In comparison to the available experimental data, this sensitivity analysis shows that the value of the force range is not a crucial quantity. It also demonstrates that a mesh width of $D/\Delta x = 20$ is adequate at the present parameter point.

Finally, let us consider the sensitivity of our results with respect to the choice of the delay time t_R. For all present simulations figure 3.4(b) shows the results for the effective coefficient of restitution computed with three different values of the delay time as indicated in figure 3.2(b): $t_R\, v_{pT}/D = 0.05$, 0.1 and 0.15. Choosing a smaller delay time has the effect of systematically increasing the coefficient of restitution, since the fluid-mediated damping after the particle-wall contact has less time to act. However, it can be observed that over the whole range of values the simulation results match the experimental data very well.

Chapter 4

Erosion of a sediment bed sheared by laminar channel flow

The main goal of this thesis is to simulate the formation of patterns from an initially flat bed of erodible sediment particles. In the previous chapter, we have presented a coupled DNS-DEM algorithm for the treatment such systems. Therein, the numerical method was validated through a standard test of a sphere colliding with a solid wall. However, the simple test case of particle-wall collision might not be a rigorous test case to ascertain the adequacy of our numerical approach to investigate dense particulate flow problems such as sedimentary pattern formation. For this purpose, we have additionally carried out extensive validation of the DEM-DNS coupling with respect to the erosion and transport of a large number of spherical particles induced by the shearing laminar flow. That is, the featureless motion of sediment particles which involves multiple contacts, sliding, rolling and saltation of particles. For this case, experimental measurements of velocity profiles of both phases are available from the recent work of Aussillous et al. (2013). The availability of detailed data as well as the direct relevance to various configurations involving shear-induced sediment erosion make the case of bedload transport in channel flow a promising candidate for a benchmark case.

We have performed 26 independent simulations covering a wide range of values of the governing parameters. The outcome of the simulations was then compared with the available experimental data with respect to the following aspects:

- the threshold for sediment erosion,
- scaling of bedload transport as a function of the principal governing parameters and
- scaling of the fluid and particle velocities both above and inside the mobile sediment bed.

Additionally, the stress balance between the particles and the fluid is assessed and compared to the prediction of a constitutive rheology model. A substantial part of the result presented in this chapter has been recently published (Kidanemariam and Uhlmann, 2014b)

4.1 Computational setup

4.1.1 Flow configuration and parameter values

We are considering the motion of spherical particles induced by laminar flow in a horizontal plane channel, as sketched in figure 4.1. The computational set-up features a number of particles, N_p in total, forming a sediment bed which takes up a large fraction of the channel height. As figure 4.1 shows, the Cartesian coordinates x, y, and z correspond to the streamwise, wall-normal and spanwise directions, respectively. The cuboidal computational domain is of size L_x, L_y, L_z in the respective coordinate directions; periodicity is imposed in x and z, while a no-slip condition is applied at the two wall planes. The flow is driven by a streamwise pressure gradient adjusted at each time step such as to maintain a constant flow rate q_f (note that the actual volumetric flow rate is divided by the spanwise domain size, i.e. q_f corresponds to a flow rate per unit span with units of velocity times length). The particulate flow problem features 10 relevant quantities (ρ_p, ρ_f, q_f, g, D, ν, L_x, L_y, L_z, N_p) which means that it is fully described by 7 non-dimensional parameters. These can be chosen as follows: the particle-to-fluid density ratio ρ_p/ρ_f; the Galileo number Ga (defined in equation (2.22)); the bulk Reynolds number given by

$$Re_b = \frac{q_f}{\nu}; \tag{4.1}$$

the global solid volume fraction defined as

$$\Phi_s = \frac{N_p V_p}{L_x L_y L_z}, \tag{4.2}$$

where $V_p \equiv (\pi/6)D^3$ is the volume of a single spherical particle; the three relative length scales L_x/D, L_y/D, L_z/D. As an alternative to one of the latter length scale ratios one can e.g. choose an imposed pile height parameter given by (Silbert et al., 2001)

$$H_b = N_p D^2/(L_x L_z). \tag{4.3}$$

Further global parameters of interest are computed a posteriori from the observed height h_f of the fluid-bed interface (defined in section 4.1.3) as well as from the characteristic shear stress. In laminar plane Poiseuille flow, the Shields number (which is defined in equation (2.21)) can be written as:

$$\Theta_{Pois} = \frac{6\,Re_b}{Ga^2}\left(\frac{D}{h_f}\right)^2. \tag{4.4}$$

Although in a region near the rough fluid-bed surface the mean velocity profile is not identical to a parabolic Poiseuille profile (cf. discussion in section 4.2.3), the Shields number based upon the definition (4.4) is often used as a parameter in experimental studies. Therefore, its value is included in table 4.1 for each simulated case. Out of the total number of N_p particles, one layer adjacent to the wall (in dense hexagonal arrangement and with small-amplitude modulation of the vertical position) is kept fixed during the entire simulation in order to form a rough bottom wall surface. The chosen parameter values in the 26 independent simulations which we have performed are listed in table 4.1. Note that 24 different physical parameter combinations have been chosen. The remaining two simulations (B20^b and B21^b) were conducted at one half the mesh width in order to verify the adequacy of the spatial resolution.

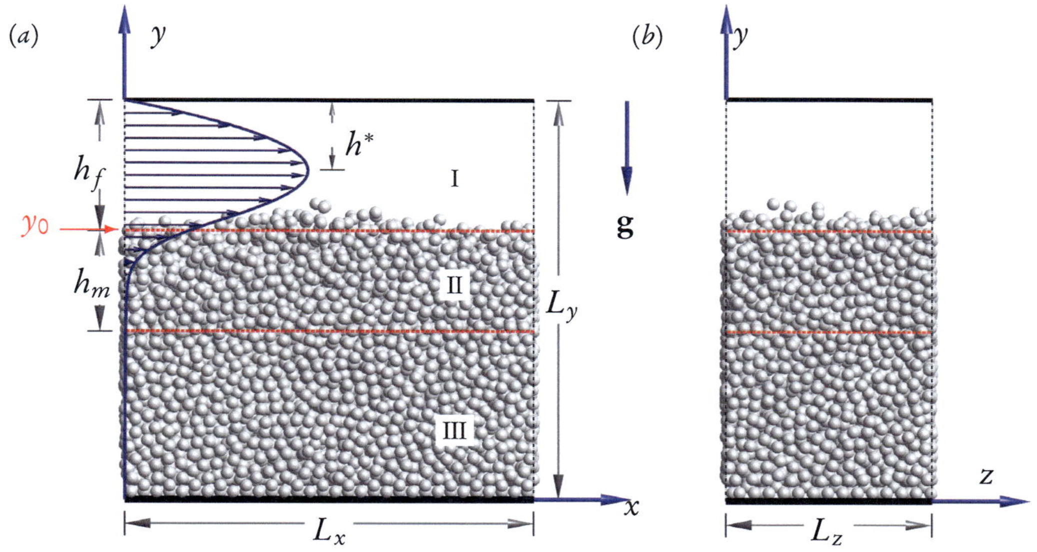

Figure 4.1: Schematic diagram showing the configuration of the bedload transport simulations. Gravity is directed in the negative y-direction. Periodicity in the x- and z-directions is assumed. (a) shows the streamwise/wall-normal plane; in (b) the view is into the streamwise direction.

Let us now turn to the parameter values of the solid contact model. Analogous to the strategy employed in section 3.2.1 we have set the stiffness parameter k_n of the elastic normal force component such that the maximum penetration length is kept below a few percent of the force range (i.e. $\max(\delta_{ij}(t)) \leq 0.05\Delta_c$). This criterion is fulfilled for values of $k_n^* = k_n D/((\rho_p - \rho_f)V_p|\mathbf{g}|)$ in the interval of $5400 - 17000$ (for the precise choice per flow case cf. table 4.1). The dry restitution coefficient was set to a value of $\varepsilon_d = 0.3$, which is smaller than the material property commonly reported for glass ($\varepsilon_d = 0.97$, cf. section 3.2.1) or plexiglass (PMMA, $\varepsilon_d = 0.8$, cf. Constantinides et al., 2008). With this choice the normal damping coefficient c_{dn} can be computed from relation (3.14). The tangential damping coefficient was set to the same value, i.e. $c_{dt} = c_{dn}$ (Tsuji et al., 1993). The Coulomb friction coefficient was fixed at $\mu_c = 0.4$ which is similar to but somewhat larger than values reported for wet contact between glass-like materials (Foerster et al., 1994; Joseph and Hunt, 2004). A sensitivity analysis of the model parameters is presented in section 4.2.1.

4.1.2 Simulation start-up and initial transient

As an initial condition for each fluid-solid simulation a sediment bed composed of quasi-randomly packed particles is generated. For this purpose, a DEM simulation was performed under the conditions of each simulation (i.e. regarding domain size, number of particles, particle diameter, particle density, value of gravitational acceleration and contact model parameters) but ignoring hydrodynamic forces, i.e. considering dry granular flow with gravity. Each DNS-DEM simulation is then started by ramping up the flow rate from zero to its prescribed value over a short time interval of the order of 10^{-2} bulk time units $T_b = h_f^2/q_f$ (where h_f is the fluid height defined in section 4.1.3). After an initial transient, a statistically stationary state of the particle-fluid system is reached with – upon average – constant bed height and constant mean particle flux. That is, initially the entire height of the channel, comprising the fluid height h_f as well as the bed height

Case	Re_b	Ga	k_n^*	H_b	Φ_s	N_p	h_f/D	Θ_{Pois}	Φ_{bed}	domain
BL01	375	6.77	13038	9.57	0.16	4900	20.04	0.12	0.44	D4
BL02	375	6.77	13038	13.03	0.21	6670	15.59	0.20	0.43	D4
BL03	375	6.77	13038	16.21	0.27	8300	10.52	0.44	0.43	D4
BL04	375	7.41	10865	20.89	0.30	10695	8.14	0.62	0.43	D5
BL05	375	8.56	8149	9.57	0.16	4900	20.16	0.08	0.44	D4
BL06	375	8.56	8149	13.03	0.21	6670	15.97	0.12	0.44	D4
BL07	375	8.56	8149	16.21	0.27	8300	11.65	0.23	0.44	D4
BL08	375	8.56	8149	18.24	0.30	9341	7.51	0.54	0.43	D4
BL09	133	13.98	5432	6.34	0.21	3246	8.03	0.06	0.44	D6
BL10	266	13.98	5432	6.34	0.21	3246	7.72	0.14	0.43	D6
BL11	333	13.98	5432	6.34	0.21	3246	7.47	0.18	0.43	D6
BL12	300	18.15	7243	6.34	0.21	3246	7.91	0.09	0.44	D6
BL13	400	18.15	7243	6.34	0.21	3246	7.78	0.12	0.44	D6
BL14	500	18.15	7243	6.34	0.21	3246	7.47	0.16	0.43	D6
BL15	111	14.86	8344	14.74	0.30	7246	8.83	0.04	0.48	D7
BL16	222	14.86	8344	14.74	0.30	7246	8.83	0.08	0.48	D7
BL17	333	14.86	8344	14.74	0.30	7246	8.63	0.12	0.48	D7
BL18	444	14.86	8344	14.74	0.30	7246	8.23	0.18	0.47	D7
BL19	200	6.30	16689	34.31	0.35	16864	12.52	0.19	0.47	D8
BL20	267	6.30	16689	34.31	0.35	16864	12.12	0.27	0.47	D8
BL21	333	6.30	16689	34.31	0.35	16864	11.42	0.39	0.47	D8
BL22	400	7.72	11126	34.31	0.35	16864	12.02	0.28	0.48	D8
BL20^b	267	6.30	16689	34.31	0.35	16864	11.96	0.28	0.47	D8^b
BL21^b	333	6.30	16689	34.31	0.35	16864	11.26	0.40	0.47	D8^b
BL23	375	7.41	10865	20.89	0.30	10695	13.90	0.21	0.53	D9
BL24	375	8.56	8149	18.24	0.30	9341	12.65	0.19	0.53	D10

Table 4.1: Physical parameters of the bedload transport simulations: bulk Reynolds number Re_b, Galileo number Ga, normalized solid stiffness coefficient $k_n^* = k_n D/((\rho_p - \rho_f)V_p|\mathbf{g}|)$, pile height parameter H_b (given in equation 4.3), global solid volume fraction Φ_s (defined in 4.2), number of particles N_p, fluid height h_f (as defined in section 4.1.3), Shields number Θ_{Pois} (defined in equation 4.4) and solid volume fraction inside the bulk of the bed Φ_{bed} (as defined in equation (4.7)). The particle-to-fluid density ratio was kept at $\rho_p/\rho_f = 2.5$ in all cases. The computational domain size and numerical parameters corresponding to the abbreviation in column 11 are listed in table 4.2.

h_b are in motion. The sediment bed slows down and a fluid velocity profile evolves over (and inside) the bed, which over time results in a statistically steady fluid velocity profile and particle flux, depending on the chosen values of the governing parameters. This way, it is insured that the steady state regime is independent of the preparation of the initial particle bed, since particle positions inside the bed are rearranged during the ramping-up stage. Animations of particle motion, which show typical time evolution of the sediment bed, are available online at http://dx.doi.org/10.1016/j.ijmultiphaseflow.2014.08.008. The time evolution of the instantaneous particle flux q_p (cf. equation (4.8) and figure 4.4, discussed below) is used to determine the statistically stationary state interval which is used to compute statistics.

Domain	$[L_x \times L_y \times L_z]/D$	$N_x \times N_y \times N_z$	$D/\Delta x$	$\Delta_c/\Delta x$	symbol
D4	$32 \times 32 \times 16$	$512 \times 513 \times 256$	16	2	—,●
D5	$32 \times 36 \times 16$	$512 \times 577 \times 256$	16	2	—,●
D6	$32 \times 16 \times 16$	$512 \times 257 \times 256$	16	2	—,●
D7	$38.4 \times 25.6 \times 12.8$	$384 \times 257 \times 128$	10	1	—,■
D8	$38.4 \times 51.2 \times 12.8$	$384 \times 513 \times 128$	10	1	—,■
D8^b	$38.4 \times 51.2 \times 12.8$	$768 \times 1025 \times 256$	20	2	—,▲
D9	$32 \times 36 \times 16$	$512 \times 577 \times 256$	16	1	—,♦
D10	$32 \times 32 \times 16$	$512 \times 513 \times 256$	16	1	—,♦

Table 4.2: Numerical parameters used in the simulations of bedload transport (for the notation cf. table 3.2). The color and symbol coding given in the last column correspond to a chosen pair of numerical parameter values $(D/\Delta x,\Delta_c/\Delta x)$ and will be used in subsequent plots.

4.1.3 Determination of the fluid height

In the reference experiments, a flourescent dye is mixed into the fluid which upon illumination by a laser sheet emits light detectable by a camera. In the recorded images the particle positions show up as low intensity (i.e. dark) regions. Such an intensity map can then be averaged over a number of frames and converted into a vertical profile of the solid volume fraction; by means of thresholding the fluid-bed interface location can be determined (Lobkovsky et al., 2008; Aussillous et al., 2013). Here we have adopted the same approach to determine the location of the interface based upon our DNS-DEM simulation data. In this manner we can directly compare our results with the experimental counterpart.

In this spirit we first define a solid phase indicator function $\phi_p(\mathbf{x},t)$ which equals unity for a point $\mathbf{x}$ that is located inside any particle at time t and zero otherwise (cf. appendix D). Next we compute the average of $\phi_p(\mathbf{x},t)$ over time (in the statistically stationary interval) and over the spanwise direction, yielding $\langle\phi_p\rangle_{zt}(x,y)$. An example of this two-dimensional map for case BL24 is shown in figure 4.2(a). Further averaging of $\langle\phi_p\rangle_{zt}(x,y)$ over the streamwise direction yields the wall-normal profile of the mean solid volume fraction $\langle\phi_p\rangle(y)$, an example of which is given in figure 4.2(b). Note that the angular brackets without subscripts $\langle\,\rangle$ denote an average over the two homogeneous space directions x, z and over time. The mean solid volume fraction $\langle\phi_p\rangle(y)$ is an alternative quantity to $\langle\phi_s\rangle(y)$ which is the same quantity but computed from the number density of particles in wall-normal averaging bins of finite size (cf. appendix D). The former approach gives more precise results in cases when there exists a strong gradient in the solid volume fraction profile as in figure 4.2; it is therefore generally preferred. The latter quantity enters the definition of the particle flux as defined in equation (4.8) below. Finally, the fluid-bed interface location is defined as the wall-normal coordinate y_0 where the value of the mean solid volume fraction equals a prescribed threshold value $\langle\phi_p\rangle^{thresh} = 0.10$, viz.

$$y_0 \quad = \quad y \ \mid \langle\phi_p\rangle(y) = \langle\phi_p\rangle^{thresh} \,. \tag{4.5}$$

Consequently, the fluid height is given by

$$h_f = L_y - y_0. \tag{4.6}$$

For the example of case BL24 the resulting interface location is shown in figure 4.2.

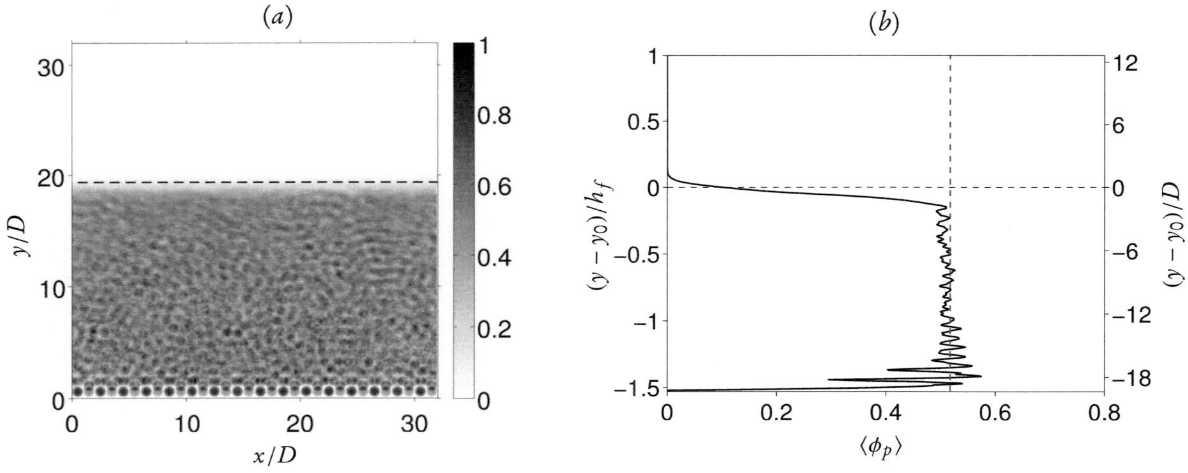

Figure 4.2: (*a*) Solid volume fraction averaged in time and over the spanwise direction, $\langle\phi_p\rangle_{zt}(x,y)$, in case BL24. The horizontal dashed line represents the mean fluid-bed interface location y_0 defined in (4.5). (*b*) The corresponding profile of the mean solid volume fraction $\langle\phi_p\rangle(y)$. The mean value in the bulk of the bed, Φ_{bed}, is indicated by a vertical dashed line (in red color).

It can also be seen in figure 4.2(*b*) that a reasonable value for the mean solid volume fraction in the bulk of the sediment bed can be defined as the following average

$$\Phi_{bed} = \frac{1}{y_2 - y_1} \int_{y_1}^{y_2} \langle\phi_p\rangle \, \mathrm{d}y \,, \tag{4.7}$$

where the interval is delimited by $y_1 = 3D$ and $y_2 = 6D$. The resulting values of Φ_{bed} vary between 0.43 and 0.53 as reported in table 4.1. Due to the combined effects of considering exactly mono-dispersed particles and using a finite force range Δ_c introduced into the collision model (cf. section 3.1.3), these values are somewhat smaller than what is found in experiments. For instance, in the study of Lobkovsky et al. (2008) values in the range of $\Phi_{bed} = 0.5 - 0.65$ are reported. Note that the maximum packing fraction of a homogeneously sheared assembly of frictional spheres measures 0.585 (Boyer et al., 2011).

4.1.4 Setup of the reference experiment

A detailed set of experimental data in a similar configuration studied by Aussillous et al. (2013) is available for validation purposes. In the experiment the transport of spherical particles in a rectangular channel in the laminar regime is considered. Two combinations of fluid and particle properties were investigated by the authors with particle-to-fluid density ratios $\rho_p/\rho_f = 1.11$ and 2.10 as well as Galileo number values $Ga = 0.36$ and 0.38. A pump is used to impose a constant flow rate, which is varied such that the bulk Reynolds number ranges from $Re_b = 0.27$ up to 1.12. Note that these values are two orders of magnitude smaller than the Reynold number values in the present simulations. However, the main control parameter of the problem (the Shields number) covers a similar range ($\Theta \approx 0.2 - 1$) as the present simulations.

Contrary to our simulation setup, the sediment bed thickness is not kept constant in the experiment. First, a given amount of sediment is filled into the channel yielding the initial sediment bed height. After start-up of the flow, particles are eroded and transported out of the test section, resulting in a fluid height which increases

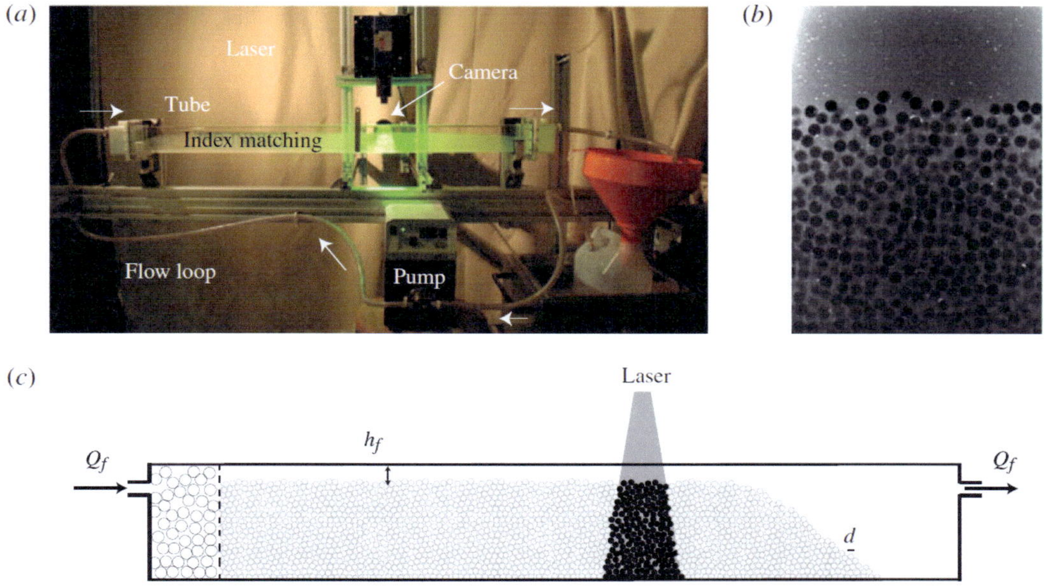

Figure 4.3: The experimental setup of Aussillous et al. (2013) (figure 1 of that publication reproduced). (*a*) Experimental apparatus. (*b*) Typical image of the bed test section. (*c*) Sketch of the setup at initial condition.

with time until it reaches its final equilibrium value. Before the cessation of particle erosion is reached, statistics are accumulated at intervals of 5 s (averaged over a small time window of 0.5 s) in which the flow is assumed to be steady. An index matching technique is utilized in order to perform optical measurements of both the velocity of the fluid phase and that of the particle phase, including locations deep inside the sediment bed. Figure 1 of Aussillous et al. (2013), which showed their experimental setup, is reproduced in figure 4.3.

4.2 Bedload particle transport by laminar channel flow

In all the simulations performed in the present work the flow remains laminar. Nevertheless, the individual particle motion is highly unsteady, exhibiting sliding, rolling, reptation, and to a lesser degree saltation, within and in the vicinity of the mobile layer of the sediment bed. It is only occasionally that particles are suspended in the flow over longer times. Therefore, transport of particles as suspended load is insignificant in the present configuration. The total rate of particle transport is dominantly bedload transport and hereafter we will not further distinguish the particle flux due to the different modes.

4.2.1 Particle volumetric flow rate

For the statistically one-dimensional flow configuration considered here, the instantaneous, total, volumetric flux of the solid phase (per unit span) is given by the following expression:

$$q_p(t) = \int_0^{L_y} \langle u_p \rangle_{xz}(y,t) \langle \phi_s \rangle_{xz}(y,t) \mathrm{d}y = \frac{\pi D^3}{6 L_x L_z} \sum_{l=1}^{N_p} u_p^{(l)}(t), \tag{4.8}$$

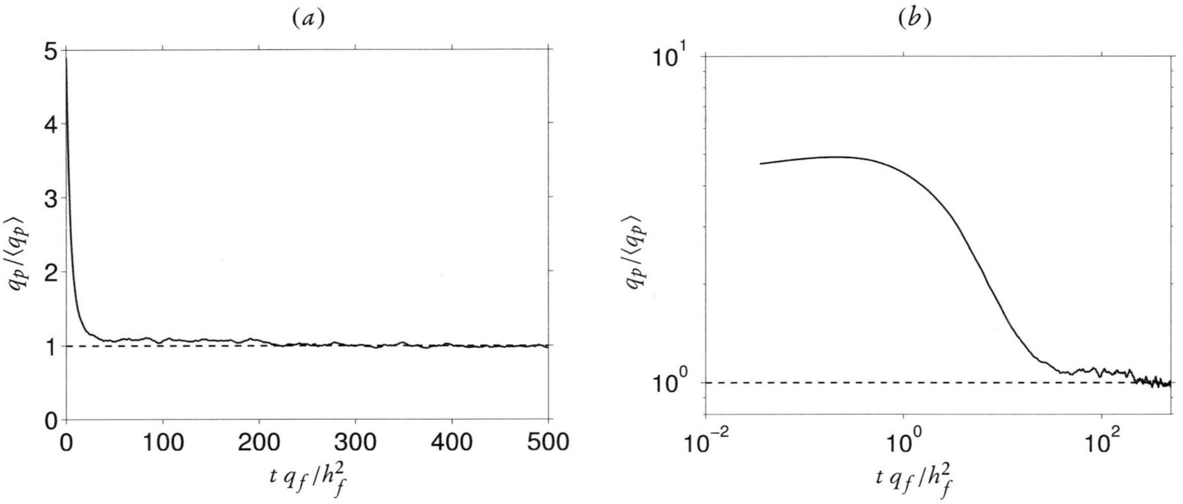

Figure 4.4: (*a*) Time evolution of the particle flow rate q_p defined in (4.8) for case BL08, normalized by the mean value in the statistically stationary regime. (*b*) shows the same data in logarithmic scale.

where $u_p^{(l)}(t)$ is the streamwise component of the instantaneous velocity of the lth particle at time t, and the operator $\langle\cdot\rangle_{xz}$ denotes an instantaneous average over the two homogeneous spatial directions (cf. appendix D). Note that the quantity q_p is termed 'grain flux' by Lobkovsky et al. (2008). The time evolution of q_p for case BL08 is shown in figure 4.4. It can be seen that after initialization of the flow the particle flow rate decreases from an initially large value, tending towards a statistically stationary state with a mean value $\langle q_p\rangle$. The duration of the transient varies among the different simulations. Since the focus of the present work is the fully developed regime, no attempt was made to find an appropriate scaling for the transient time.

Figure 4.5(*a*) shows the mean particle flow rate in the statistically stationary regime, $\langle q_p\rangle$, plotted for all cases as a function of the Shields number Θ. The experimentally determined critical Shields number in the laminar flow regime, $\Theta_c = 0.12$, together with the confidence interval ± 0.03 (Ouriemi et al., 2007) are indicated with vertical lines in figure 4.5(*a*). One possible reference quantity for the particle flow rate is the following viscous scale as suggested by Aussillous et al. (2013),

$$q_v = \frac{(\rho_p/\rho_f - 1)|\mathbf{g}|D^3}{\nu} = Ga^2\,\nu\,, \tag{4.9}$$

which is equivalent to $18U_sD$, where U_s is the Stokes settling velocity of a single particle. The quantity defined in equation (4.9) is used to non-dimensionalize $\langle q_p\rangle$ in figure 4.5(*a*), while an alternative normalization will be explored in figure 4.5(*b*). It is clearly observable from figure 4.5(*a*) that the particle flow rate obtained from the present DNS-DEM is negligibly small for values of $\Theta_{Pois} \lesssim 0.12$ in good agreement with what is reported in the literature (cf. Charru et al., 2004; Loiseleux et al., 2005; Ouriemi et al., 2009b). Animations of the particle motion (as provided online at http://dx.doi.org/10.1016/j.ijmultiphaseflow.2014.08.008) confirm this finding. Those movies also show the well-known intermittent movement of a small number of particles for Shields number values smaller than Θ^c_{Pois}, as has been observed experimentally (Charru et al., 2004).

On the other hand, in those runs in which the value of the Shields number is above the critical value, the particles near the fluid-bed interface are observed to be continuously in motion. The magnitude of the saturated particle flux $\langle q_p\rangle$ monotonically increases with increasing value of Θ_{Pois}. This trend is clearly seen

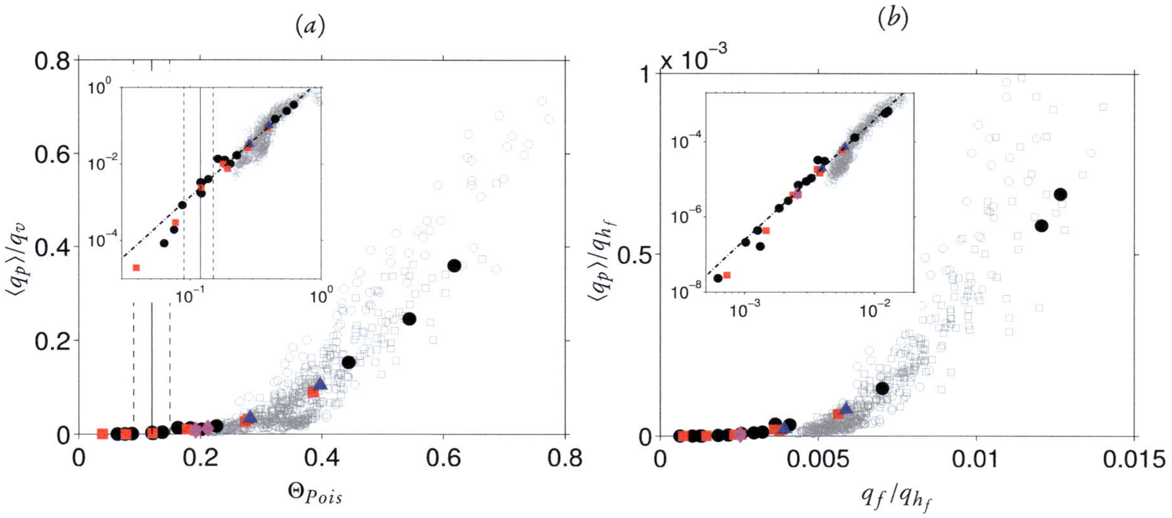

Figure 4.5: (*a*) Mean particle flow rate $\langle q_p \rangle$ non-dimensionalized by the viscous scaling q_v plotted as a function of the Shields number Θ_{Pois}. The vertical solid line corresponds to the critical Shields number for particle erosion $\Theta^c_{Pois} = 0.12 \pm 0.03$ reported by Ouriemi et al. (2007), the dashed lines indicating the tolerance range. The inset shows the same data plotted in logarithmic scale. The black chain-dotted line in the inset corresponds to a power law $\langle q_p \rangle / q_v = 1.6584\Theta^{3.08}_{Pois}$. (*b*) The same quantity $\langle q_p \rangle$ as in (*a*), but normalized by q_{h_f} and plotted as a function of the non-dimensional fluid flow rate q_f / q_{h_f}. The power law $\langle q_p \rangle / q_{h_f} = 1055.3(q_f / q_{h_f})^{3.21}$ is indicated by the black chain-dotted line. In both graphs the results from the present simulations are indicated by the symbols given in table 4.2. Data points marked by (○,□) correspond to the experiment of Aussillous et al. (2013) (○, combination "A" of materials in their table 1; □, combination "B").

in figure 4.5(*a*). The corresponding experimental data of Aussillous et al. (2013) is also added to the figure, exhibiting a very good agreement with our present DNS-DEM results. It should be noted that the particle flow rate defined in equation (4.8) is not presented in the paper by Aussillous et al. (2013). We have computed it from the data provided by the authors as supplementary material. Moreover, a fit of the simulation data in figure 4.5(*a*) for super-critical parameter points ($\Theta_{Pois} > \Theta^c_{Pois}$) yields the following power law relation:

$$\langle q_p \rangle / q_v = 1.6584\Theta^{3.08}_{Pois}. \tag{4.10}$$

Ouriemi et al. (2009b) have described the present bedload transport system by means of a two-fluid model which makes use of the Einstein effective viscosity. They have shown that above the critical value Θ^c_{Pois} the model predicts a variation of the particle flow rate with the third power of the Shields number. As can be seen from the fit in figure 4.5(*a*), their conclusion is fully confirmed by the present data.

An alternative scaling of the particle flow rate is to replace the diameter D in (4.9) with the fluid height:

$$q_{h_f} = \frac{(\rho_p/\rho_f - 1)|\mathbf{g}|h_f^3}{\nu} = q_v \left(\frac{h_f}{D}\right)^3 . \tag{4.11}$$

This scaling was advocated by Aussillous et al. (2013) as more consistent with a continuum (two-fluid) approach. Note that this scaling does not take into account the finite size of the particles. Figure 4.5(*b*) shows that our data for the particle flow rate as a function of the fluid flow rate, both normalized with the scale defined in equation (4.11), again matches the experimental data of Aussillous et al. (2013) very well. We also

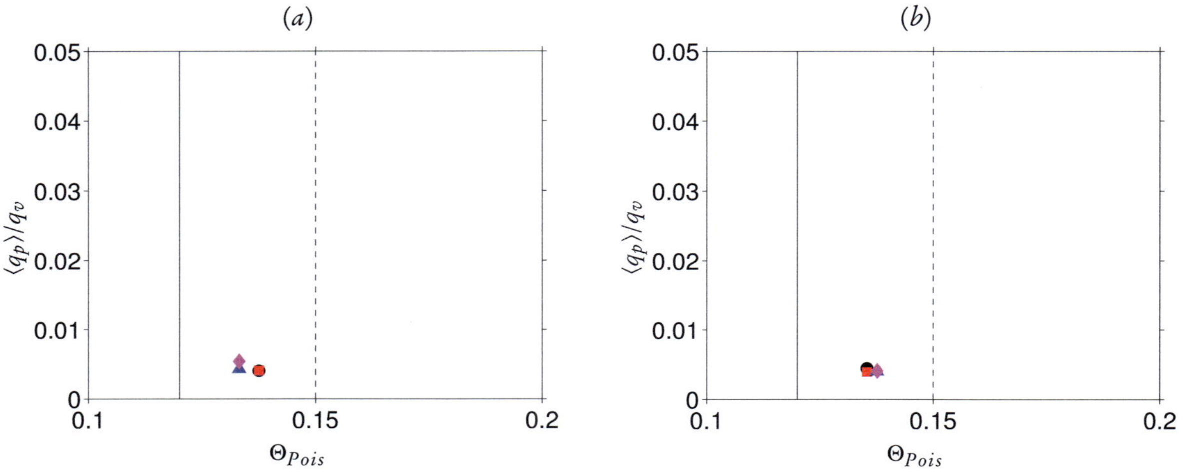

Figure 4.6: The same plot as figure 4.5 but showing the effect of changing the parameters of the collision model in case BL10 (cf. table 4.1). In (*a*) the dry restitution coefficient ε_d is varied while all remaining physical and numerical parameter values are kept the same. ●, $\varepsilon_d = 0.30$; ■, $\varepsilon_d = 0.60$; ▲, $\varepsilon_d = 0.90$; ◆, $\varepsilon_d = 0.97$. In (*b*) only the Coulomb friction coefficient μ_c is varied. ●, $\mu_c = 0.10$; ■, $\mu_c = 0.25$; ▲, $\mu_c = 0.40$; ◆, $\mu_c = 0.55$. Note that the present axis scales are different than in figure 4.5.

note that the simulation data is very well represented by the power law:

$$\langle q_p \rangle / q_{h_f} = 1055.3 (q_f / q_{h_f})^{3.21} \tag{4.12}$$

which is again close to a cubic variation. In fact it can be shown that a cubic variation of $\langle q_p \rangle / q_{h_f}$ with q_f / q_{h_f} follows directly from a cubic variation of $\langle q_p \rangle / q_v$ with Θ_{Pois}.

Sensitivity with respect to collision model parameters

The force range Δ_c has been varied by a factor of two (from $\Delta_c = \Delta x$ to $\Delta_c = 2\Delta x$) in two simulations (BL20 and BL21, cf. tables 4.1 and 4.2). It can be seen from figure 4.5 that the effect of modifying this parameter upon the particle flow rate is insignificant in the given range. We have likewise tested the influence of the choice of the value for the dry restitution coefficient ε_d upon the results in the present configuration. For this purpose the flow case dented BL10 in table 4.1 has been repeated three times with modified values of the dry restitution coefficient (while maintaining all remaining numerical and physical parameters at their original value). Figure 4.6(*a*) shows the resulting mean particle flow rate for the set of coefficients $\varepsilon_d = 0.3, 0.6, 0.9, 0.97$. As can be observed, the impact of this parameter variation is very limited. Finally, we have repeated the same case BL10 while changing the value of the Coulomb friction coefficient in the range $\mu_c = 0.1, 0.25, 0.4, 0.55$. The result is shown in figure 4.6(*b*); again a very small effect on the mean particle flow rate is obtained.

4.2.2 Thickness of the mobile sediment bed

In practical applications involving bedload transport it is often of interest to predict the extent of the layer of particles which exhibits significant streamwise motion. Moreover, from the point of view of performing bedload transport simulations, having an estimation of the mobile sediment bed layer, prior to carrying out the simulations, is important. The thickness of the mobile sediment bed layer determines the extent of the total sediment bed height which should be imposed to guaranty a faithful simulation prediction. Referring to

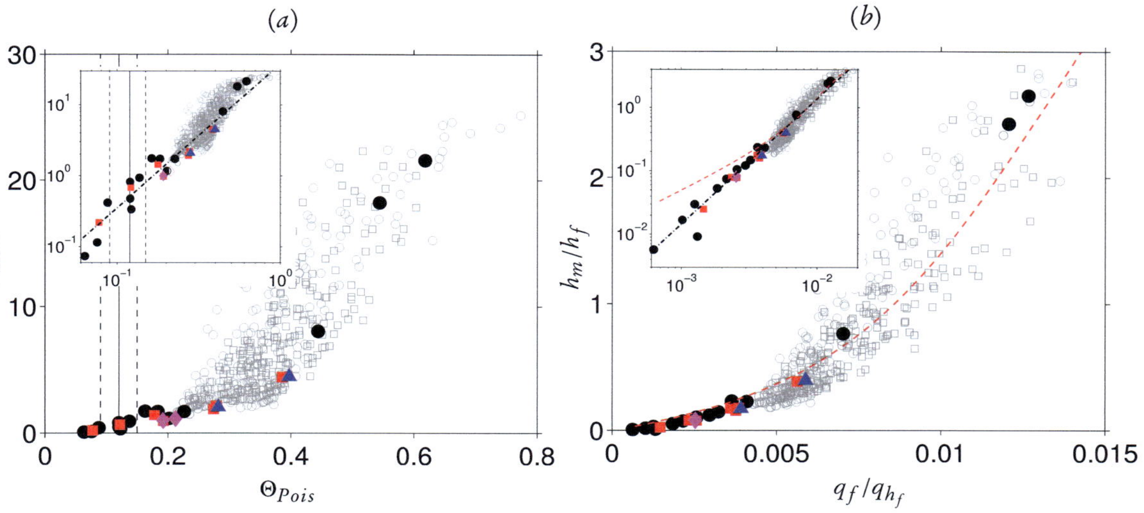

Figure 4.7: Thickness of the mobile bed layer h_m: (*a*) normalized by the particle diameter D and given as a function of the Shields number Θ_{Pois}; (*b*) normalized by the fluid height h_f and given as a function of the non-dimensional fluid flux q_f/q_{h_f}. The insets show the same data in logarithmic scale. The chain-dotted lines indicate the following data fits: (*a*) $h_m/D = 38.01\,\Theta^{2.03}$; (*b*) $h_m/h_f = 12096\,(q_f/q_{h_f})^{1.97}$. Data points marked by (○,□) correspond to the experiment of Aussillous et al. (2013).

the schematic of the configuration shown in figure 4.1, we define the thickness of the mobile layer h_m as the distance between the fluid-bed interface location (y_0) and the location inside the bed where the mean particle velocity $\langle u_p \rangle$ is equal to a prescribed threshold value $\langle u_p \rangle^{thresh}$. In the present work we set this threshold velocity to $\langle u_p \rangle^{thresh} = 0.005 \max(\langle u_f \rangle)$. In terms of the Stokes settling velocity, the chosen value ranges from $0.01U_s$ to $0.09U_s$ in the different cases which we have simulated. In the study of Aussillous et al. (2013) the mobile layer thickness h_m has been determined by a similar thresholding criterion.

Figure 4.7(*a*) shows the computed mobile layer thickness h_m normalized by the diameter of the particles as a function of the Shields number; figure 4.7(*b*) shows the same quantity under the alternative normalization with the fluid height, plotted as a function of the non-dimensional fluid flow rate. Once again, the DNS data shows very good agreement with the experimental data despite the sensitivity of this quantity due to thresholding. The data in figure 4.7 clearly shows that h_m monotonically increases with increasing values of the Shields number and with the non-dimensional fluid flow rate. It can be observed that the DNS-DEM data over the presently investigated parameter range is fairly well represented by a quadratic law in both graphs.

The quadratic variation observed from the simulations is in contrast to the available models in the literature. Mouilleron et al. (2009) have considered the viscous re-suspension theory of Leighton and Acrivos (1986) in which particles are assumed to have no inertia and a mass balance between downward sedimentation and upward diffusion is considered. Ouriemi et al. (2009b) on the other hand consider a continuum description of the problem where a frictional rheology is assumed to describe the mobile granular layer. In both theoretical approaches (Mouilleron et al., 2009; Ouriemi et al., 2009b) the authors arrive at a linear variation of the thickness of the mobile layer with the Shields number. Please note, however, that the definition of the mobile layer thickness in the continuum model context is not identical to the definition which is employed in both the present work and in the experiment (Aussillous et al., 2013). Going back to the present simulation data, it can be seen that at larger values of the two alternative control parameters our results are in fact not inconsistent with a linear variation of the mobile layer thickness with both Shields number (figure 4.7(*a*)) and with the fluid flow rate (figure 4.7*b*). However, in view of the fact that the data points (both in the experiment and in the simulation) cover only a limited range of these control parameters, and considering the scatter of the experimental data, this issue cannot be settled at the present time.

4.2.3 Fluid and particle velocities

The wall-normal profiles of the streamwise component of the mean fluid and particle velocities $\langle u_f \rangle$ and $\langle u_p \rangle$ for all cases are shown in figure 4.8. In this figure the length and velocity scales proposed by Aussillous et al. (2013) (h_f and q_{h_f}/h_f, respectively) are used. The graph in figure 4.8(*a*) features an ordinate which is shifted to the fluid-bed interface location (y_0), while in figure 4.8(*b*) the location of the bottom of the mobile layer ($y_0 - h_m$) is used as the zero of the ordinate. Note that, velocities corresponding to the suspended particles located in the averaging bins above the bed-fluid interface, are not shown in the figures for clarity. Similar to what has been observed in the experiments (Aussillous et al., 2013), the present profiles exhibit three distinct regions: (I) the clear fluid region ($0 < y - y_0 < h_f$), where the mean fluid velocity profile is characterized by a near-parabolic shape, while the solid volume fraction is negligibly small; (II) the mobile granular layer ($-h_m < y - y_0 < 0$), where both the fluid and particles are in motion; (III) the bottom region ($y < y_0 - h_m$), where the velocities of both phases are vanishingly small. It can be observed from figure 4.8 that there exists no significant difference between the mean velocities of the two phases. This finding is also in agreement with previous the observation made by (Aussillous et al., 2013). Note that in the clear fluid region (I) particles are occasionally entrained into the bulk flow, reaching larger distances above the mobile layer. However, the mean particle velocity at these points was not determined with sufficient statistical accuracy, and these values are, therefore, not shown in figure 4.8. In the follwing, we characterize the two regions, i.e. the clear fluid region and the mobile sediment bed layer region.

Clear fluid region

In the laminar flow regime the fluid velocity profile in the clear fluid region of the channel would be exactly of parabolic shape if that region were strictly devoid of particles. Deviations from the parabola are investigated in figure 4.9. For this purpose we normalize the mean particle velocity $\langle u_f \rangle(y)$ by its respective maximum value $U_f^m = \max\langle u_f \rangle$, and define a coordinate measuring the distance from the top wall, $\tilde{y} = y - L_y$. We also denote by h^* the distance of the location of U_f^m measured from the top wall (cf. the schematic diagram in figure 4.1). Figure 4.9(*a*) shows a graph of $\langle u_f \rangle$ under this scaling. It can be seen that all cases collapse upon the Poiseuille flow parabola,

$$\frac{\langle u_f \rangle}{U_f^m} = -\left(\frac{\tilde{y}}{h^*}\right)^2 - 2\left(\frac{\tilde{y}}{h^*}\right), \tag{4.13}$$

over a large wall-normal interval. Deviations from the parabolic shape are noticeable for $y \leq y_0$, which is a case dependent location under the scaling of figure 4.9(*a*). These deviations are of the form of larger velocity values than those given by (4.13) around the fluid-bed interface. Note that in a smooth wall laminar Poiseuille flow we have $h^* = h_f/2$ and $q_f = 2U_f^m h_f/3$. Thus the dependence of the characteristics of the fluid velocity profile on the non-dimensional fluid flow rate can be inferred by examining the variation of $U_f^m h_f/q_{h_f}$ and h^*/h_f as a function of q_f/q_{h_f}. These quantities are shown in figure 4.9(*b*)and 4.9(*c*). It is seen that, for values of q_f/q_{h_f} smaller than approximately 0.004, the profiles seem to be adequately described by smooth wall Poiseuille flow. At larger values of the non-dimensional flow rate, however, U_f^m is observed to deviate progressively from the Poiseuille values, exhibiting an increasingly smaller slope. The opposite trend is found for h^*, which grows beyond the Poiseuille flow value with increasing q_f/q_{h_f}. Figure 4.9(*b*,*c*) also includes

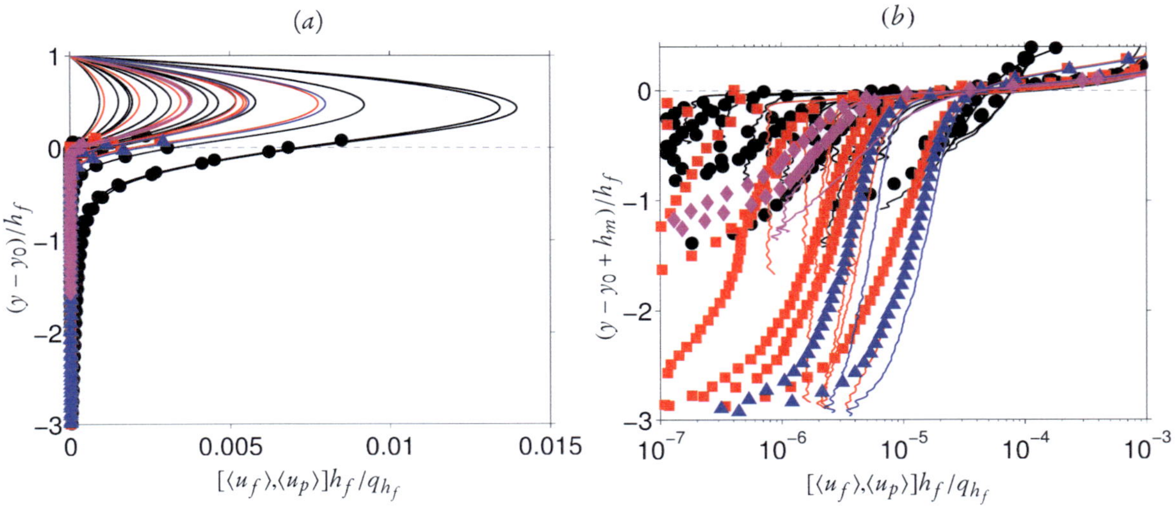

Figure 4.8: Wall-normal profiles of the streamwise component of the mean fluid velocity $\langle u_f \rangle$ (solid lines) and of the mean particle velocity $\langle u_p \rangle$ (symbols); color and symbol coding as indicated in table 4.2. In (*a*) the ordinate is shifted to the fluid-bed interface location, while in (*b*) it is shifted to the lower boundary of the mobile layer.

corresponding values computed from the available experimental data of Aussillous et al. (2013). Despite some scatter in the values from the experiments, a very good match with the simulation data is obtained for U_f^m. On the other hand, the experimental values of h^* are highly scattered, making a direct comparison difficult.

Mobile granular layer region

In this region, the particles and the fluid are observed to move at almost the same mean velocity. Thus addressing only the fluid streamwise velocity $\langle u_f \rangle$ should suffice. In order to finding an appropriate scaling of the velocity profile, we chose the maximum value of the fluid velocity in this region, i.e. the value at the fluid-bed interface $y = y_0$, as a velocity scale (denoted as U_f^0). For the length scale, the thickness of the mobile layer h_m was initially adopted. However, it turns out that this length scale does not properly scale the profiles (plots not shown). This could be related to the uncertainties inherited in determining h_m, since the interface between the mobile and stationary sediment layers is not clearly defined (cf. the discussion in section 4.2.2). As is shown in figure 4.8, the velocity profiles are observed to sharply fall from the value U_f^0 at the fluid-bed interface down to a certain penetration into the sediment bed. This interval, where the mean velocity profiles of all cases exhibit a large wall-normal gradient, is smaller in thickness when compared to the remaining interval, where the velocity is observed to gradually vanish to zero. Thus, it makes sense to define a length scale as a wall-normal distance from the fluid-bed interface down to a wall-normal location where $\langle u_f \rangle$ equals a certain fraction of U_f^0. The amount of the fraction is chosen such that the location falls within the interval of rapid wall-normal change. For convenience, we choose the location at which $\langle u_f \rangle$ attains a value which is 50% of the values at the fluid-bed interface. Hereafter, this length scale is termed as the 'characteristic half thickness' of the mobile bed, $\tilde{h}$, and thus, is given viz.

$$\tilde{h} = y \mid \langle u_f \rangle(y) = 0.5 U_f^0 \quad \forall y \in [y_0 - h_m, y_0] \,. \tag{4.14}$$

Such a choice is motivated by the classical 'half-width' of a free round jet flow (cf. Pope, 2000, page 97) and allows an unambiguous way of defining an appropriate length scale.

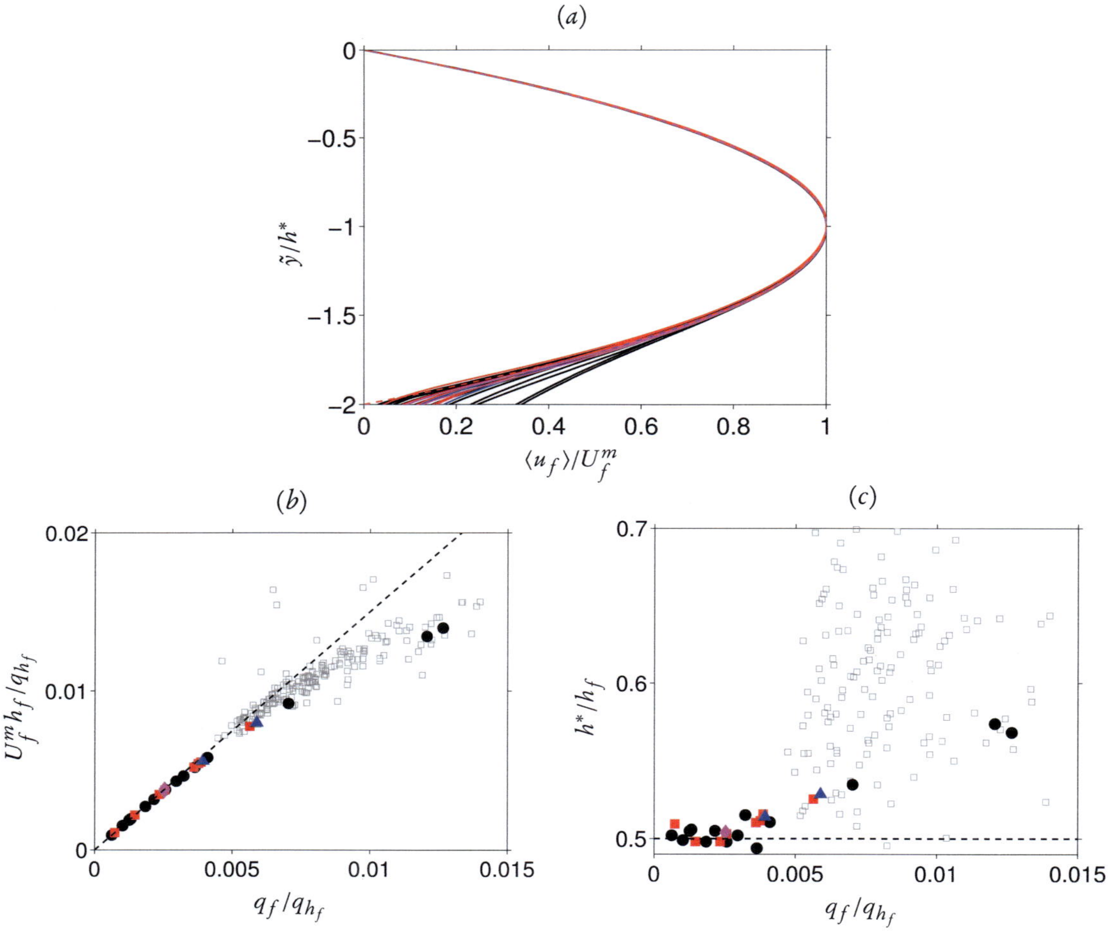

Figure 4.9: (*a*) Wall-normal profile of the fluid velocity $\langle u_f \rangle$ normalized by its maximum value U_f^m. The vertical axis corresponds to $\tilde{y} = y - L_y$, scaled by the wall-normal distance h^* between the top upper wall and the location of the maximum velocity U_f^m. The red dashed curve corresponds to the Poiseuille flow parabola (4.13). (*b*) The variation of U_f^m as a function of q_f. The dashed line corresponds to the value for smooth wall Poiseulle flow, i.e. $(3/2)q_f/h_f$. (*c*) The variation of h^*as a function of q_f with the Pouiseuille value $h^*/h_f = 1/2$ marked by a dashed line. Data points (○,□) correspond to the experiment of Aussillous et al. (2013).

Figure 4.10(*a*) shows the mean fluid velocity profiles, non-dimensionalized with the above mentioned scales, for all the cases. The profiles are observed to satisfactorily collapse showing the relevance of the above defined length scale. The scaling of the fluid (and particle) velocity in this region can now be inferred by looking into the variation of U_f^0 and $\tilde{h}$ as a function of Θ_{Pois} or q_{h_f}. These quantities are shown in figures 4.10(*b–e*). It can be observed that both quantities increase monotonically as a function of the Shields number (or the non-dimensional flow rate) indicating that the fluid velocity inside the sediment bed is sufficiently described by the Shields number (or the non-dimensional flow rate). It seems that a the fluid velocity at the fluid-bed interface scales approximately quadratically with the Shields number, or the non-dimensional flow rate, at least in the considered parameter span. On the other hand, the characteristic half thickness is observe to scale approximately linearly with the Shields number. The respective power law fits are provided in figures 4.10(*b–e*)). Corresponding data from the reference experiment is also shown in the figures. Although it exhibits a scatter, the experimental data is seen to be in good agreement with the DNS.

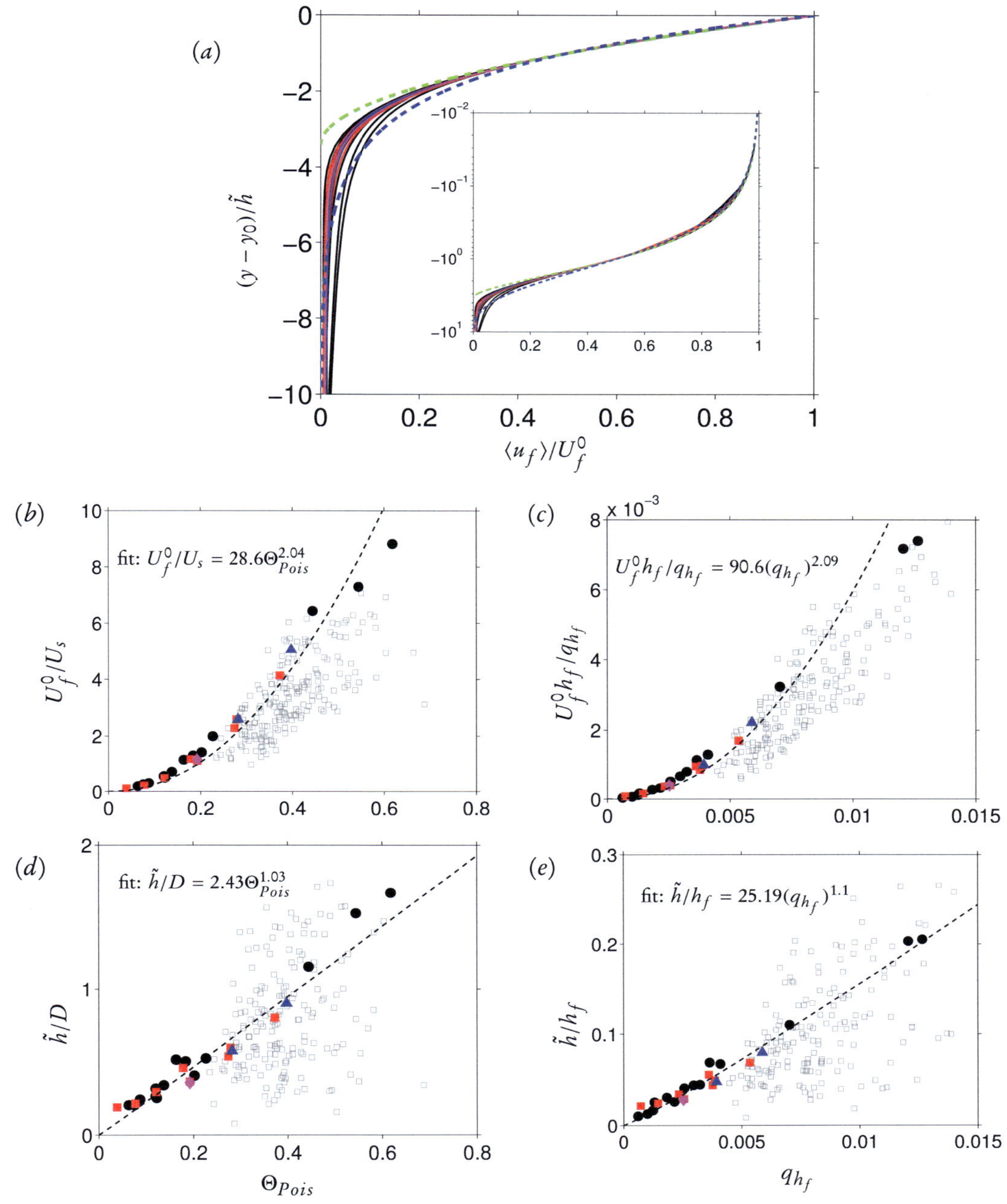

Figure 4.10: (*a*) Wall-normal profile of mean streamwise fluid velocity in the granular mobile layer region, normalized by its value at the fluid-bed interface, for all the cases. The wall-normal axis is scaled with the characteristic half thickness of the mobile bed defined in equation (4.14). The green dashed curve (---) corresponds to a parabolic model defined in equation (4.15), while the blue dashed curve (---) corresponds to the exponential model defined in equation (4.16). The Mean fluid velocity at the interface U_f^0, normalized by U_s, as a function of the Shields number is shown in (*b*). The same data, normalized by the scaling proposed by Aussillous et al. (2013), as a function of the non-dimensional fluid flow rate is shown in (*c*). (*d*) shows the characteristic half thickness $\tilde{h}$, normalized by the particle diameter, as a function of the Shields number, while (*e*) shows $\tilde{h}$ normalized by the fluid height as a function of the nondimensional flow rate. Data points marked by (□) correspond to the experiment of Aussillous et al. (2013).

Some previous studies have proposed a parabolic profile of the mean particle velocity inside the bed (Mouilleron et al., 2009; Ouriemi et al., 2009b; Aussillous et al., 2013). To see if the DNS data exhibits a parabolic trend, we define a parabola fully characterized by U_p^0 and $\tilde{h}$ as

$$\frac{\langle u_f \rangle}{U_f^0} = \left(\frac{y - y_0}{\widetilde{H}}\right)^2 + 2\left(\frac{y - y_0}{\widetilde{H}}\right) + 1 \qquad \forall y \in [y_0 - \widetilde{H}, y_0] \tag{4.15}$$

where $\widetilde{H} \equiv (2 + \sqrt{2})\,\tilde{h}$ corresponds to the wall-normal location at which $\langle u_f \rangle = 0$. The parabola (4.15) is compared to the DNS data in figure 4.10(a). In the wall-normal range $-2\tilde{h} \lesssim y - y_0 \leq 0$, it can be seen that the model (4.15) matches very well with the DNS data. For $y - y_0 \lesssim -2\tilde{h}$ on the other hand, the parabola is seen to deviate from the DNS data appreciably, underestimating $\langle u_f \rangle$. That is, while the parabola predicts $\langle u_f \rangle = 0$ at $y = -\widetilde{H}$, the DNS data exhibits a gradual decrease of $\langle u_f \rangle$ and attains small but finite values of streamwise velocity even down to $y \approx -10\tilde{h}$. This indicates that the velocity profiles deep inside the bed can not be entirely represented by a parabolic fit.

Recently, Hong et al. (2015) have fitted an exponential curve to the experimental data of Goharzadeh et al. (2005). The latter have investigated the flow velocity at the interface between a stationary porous media (composed of spherical particles) and an overlying fluid. In the same spirit, we define an exponential model which is fully described by U_p^0 and $\tilde{h}$ as

$$\frac{\langle u_f \rangle}{U_p^0} = \exp\left\{-\log(0.5)\left(\frac{y - y_0}{\tilde{h}}\right)\right\}, \quad y < y_0 \,. \tag{4.16}$$

Figure 4.10(a) also shows the exponential model (4.16) plotted together with the DNS data. Similar as the parabolic model, the exponential curve sufficiently describes the DNS data in the wall-normal range $-2\tilde{h} \lesssim y - y_0 \leq 0$. Deep inside the sediment bed (for $y - y_0 \lesssim -2\tilde{h}$), the model is seen to slightly overestimate the velocity profiles. Nevertheless, owing to the slowly decaying nature of the exponential curve at large wall-normal distances from the fluid-bed interface, it can be argued that the velocity profile in this region is better represented by equation (4.16) rather than the parabolic one. The physical relevance of an exponential model to the velocity profile inside the granular layer is something which needs further investigation.

4.2.4 Momentum balance: Hydrodynamic and collision forces

Forces acting on individual particles

In the present configurations, particle motion is driven by hydrodynamic forces, buoyancy as well as contact forces. Let us denote the component of the mean hydrodynamic force acting on particles in the ith direction as $\langle F_{p,i}^H \rangle$. Likewise, the mean inter-particle collision forces are denoted $\langle F_{p,i}^C \rangle$. Moreover, the mean total force, which is the sum of the hydrodynamic and collision part, is denoted as $\langle F_{p,i} \rangle$. Note that in the wall-normal direction, the total force has an additional contribution from the submerged weight of the particle:

$$F_W = (\rho_p - \rho_f) V_p |\mathbf{g}| \tag{4.17}$$

which is acting in the negative wall-normal direction. In equation (4.17), $V_p = \pi D^3/6$ is the volume of a single particle and $|\mathbf{g}|$ is the magnitude of acceleration due to gravity. The tangential components of the hydrodynamic and collision forces acting on the particles are responsible for the observed rotational motion

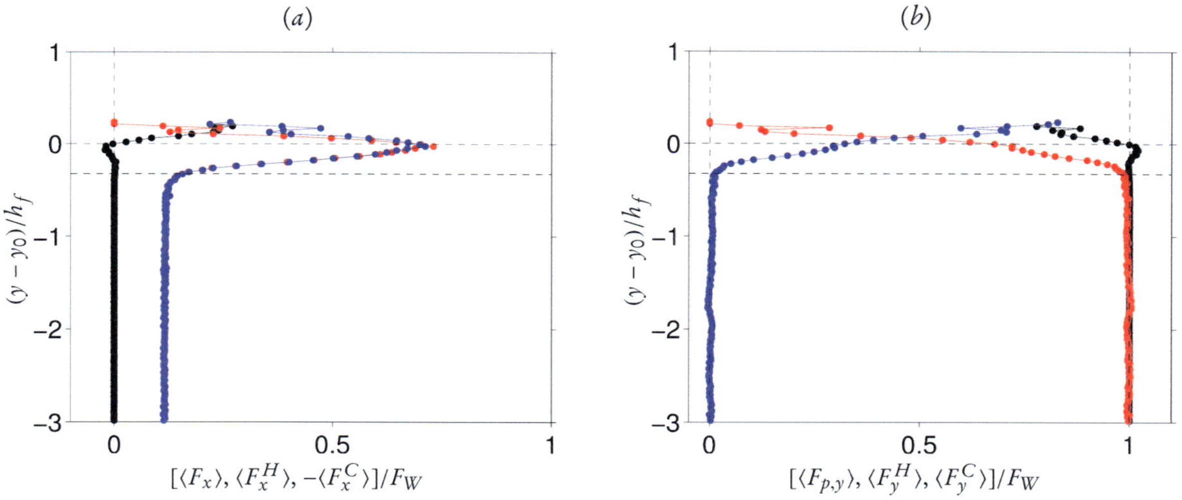

Figure 4.11: Wall-normal profile of the mean hydrodynamic and collision forces acting on particles corresponding to case BL21; (*a*) streamwise component; —•—,$\langle F_x^H \rangle$; —•—,$\langle F_x^C \rangle$; —•—,$\langle F_x \rangle = \langle F_x^H \rangle + \langle F_x^C \rangle$. (*b*) wall-normal component. —•—,$\langle F_y^H \rangle$; —•—,$\langle F_y^C \rangle$; —•—,$\langle F_{p,y} \rangle$. In both plots, the horizontal dashed line indicate the location of the fluid-bed interface and the extent of the mobile sediment bed layer. The submerged weight F_W (4.17) is used for normalization.

of particles. In the quest of understanding particle dynamics, characterization of the rotational part of the particle motion is equally important as the translational part. For instance, our previous studies have shown that the mean particle rotation rate is directly proportional to the mean shear rate of the flow (García-Villalba et al., 2012; Kidanemariam et al., 2013). However, such an analysis is not included in the present work and deserves detailed investigation in the future.

Figure 4.11(*a*) shows a sample wall-normal profile of $\langle F_x^H \rangle$ and $\langle F_x^C \rangle$ corresponding to case BL21. All the other cases exhibit similar wall-normal variation and thus their corresponding plots are not shown. As expected, in the stationary sediment bed region, the net force acting on the particles is practically zero. In this region, it is seen that collision forces counter balance hydrodynamic forces. The hydrodynamic force which the particles feel in this region is expected to be the forcing due to the driving pressure gradient, i.e. $\langle F_x^H \rangle \propto \langle dp/dx \rangle$. Owing to the small velocities, Darcy's drag is negligible. In the mobile bed region, the hydrodynamic force acting on the particles increases monotonically from its value inside the stationary bed region to its maximum value near the fluid-bed interface. The collision force also exhibits similar trend but counterbalancing the hydrodynamic force. Thus, similar as in the stationary bed, in most part of the mobile granular layer, the mean streamwise total force is seen to be negligibly small. However, we observe a small but finite mean deceleration of particles ($\langle F_x^H \rangle < \langle F_x^C \rangle$) close the fluid-bed interface. On the contrary, a substantial mean acceleration of particles ($\langle F_x^H \rangle > \langle F_x^C \rangle$) is observed in region above the fluid-bed interface. Although the flow is statistically steady, the mean force acting on the particles in the mobile sediment bed layer as well as in the clear fluid region close to the fluid-bed interface does not need to be zero [1]. This is believed to be related to the fluctuating wall-normal motion of particles. As is observed in the previously mentioned animations of particle motion, near the fluid-bed interface particles move not in a statistically horizontal line rather in a periodic cycle of erosion-deposition which leads to the observed acceleration/deceleration regions.

[1] A non-zero mean acceleration of fluid and finite-size particles has also been observed in a turbulent plane channel flow (García-Villalba et al., 2012)

As is shown in figure 4.11(*b*) the mean hydrodynamic force in the wall-normal direction inside the stationary bed is, as expected, vanishingly small. The submerged weight of the particles is balance by the wall-normal component of the collision forces. In the mobile sediment layer on the other hand, the hydrodynamic contribution to the total mean force (lift) monotonically increases with increasing wall-normal location while the mean collision force decreases. Consequently, the mean total force is almost zero (after subtracting the submerged weight of the particles) in all wall-normal locations except for regions near the fluid bed interface. The non-zero mean wall-normal force could be explained by similar arguments as in streamwise direction.

Streamwise momentum balance

It has been stated in section 4.1.1, that the pressure gradient is adjusted at each time step in order to maintain the imposed constant flow rate q_f of each simulation. Thus the value of the pressure gradient is not known *a priori*. However, it can be determined from the balance between the driving volume force and resisting forces. The resisting forces comprise shear resistance from the smooth wall located at the top of the domain and stationary particles located at the bottom. The contribution from the bottom smooth wall beneath the stationary particles is negligible due to the almost zero velocity gradients therein. Thus,

$$\frac{\mathrm{d}p}{\mathrm{d}x}(t) = -\frac{1}{L_x L_y L_z}\sum_{l=1}^{N_p}\left(F_{p,x}^{H(l)}(t) + F_{p,x}^{C(l)}(t)\right)I_{stat}^{(l)} - \frac{1}{L_y}\langle\tau^W\rangle_{xz}(t)\,, \tag{4.18}$$

where $I_{stat}^{(l)}$ is an indicator function which has a value of unity if the lth particle is stationary and zero otherwise. $\langle\tau^W\rangle_{xz}$ is the instantaneous average shear stress at the top smooth wall. Figure 4.12(*a*) shows the time evolution of the pressure gradient and the different counterbalancing terms in equation (4.18) for case BL21. It can be seen that the pressure gradient decreases from an initial large value towards its steady-state value $\langle \mathrm{d}p/\mathrm{d}x\rangle$. This trend is associated with the initialization conditions of the simulations in which at the beginning, the whole bed and fluid are set in motion. The figure also shows, that the pressure gradient is dominantly counterbalanced by the collision forces between the mobile and stationary particles.

At any given wall-normal location, the total mean shear stress $\langle\tau\rangle$ is the sum of the fluid viscous shear stress and shear stress contribution stemming from the fluid-particle interaction and balances the imposed mean pressure gradient. In the steady-state regime, the wall-normal profile of $\langle\tau\rangle$ is a straight line with a slope given by the value of the mean pressure gradient:

$$\langle\tau\rangle(y) = \langle\tau_f\rangle(y) + \langle\tau_p\rangle(y) = \langle\tau^W\rangle - \langle \mathrm{d}p/\mathrm{d}x\rangle(L_y - y) \tag{4.19}$$

In equation (4.19), the viscous fluid shear stress of the pure fluid is given as:

$$\langle\tau_f\rangle(y) = \rho_f \nu \dot{\gamma}_f(y) = \rho_f \nu \frac{\mathrm{d}\langle u_f\rangle}{\mathrm{d}y} \tag{4.20}$$

where $\dot{\gamma}_f$ is the fluid shear rate. In the context of the immersed boundary method, the Navier-Stokes equations are solved on the entire domain comprising the fluid domain and the space occupied by the suspended particles. Thus the particles exert a volume force $\mathbf{f} = (f_x, f_y, f_z)^T$ which imposes the no-slip condition at the fluid-particle interface (cf. chapter 3.1.2). The particle related shear stress is expressed in terms of this volume force as (see e.g. Kidanemariam et al., 2013):

$$\langle\tau_p\rangle(y) = -\rho_f \int_{L_y}^{y} \langle f_x\rangle(y)\mathrm{d}y \tag{4.21}$$

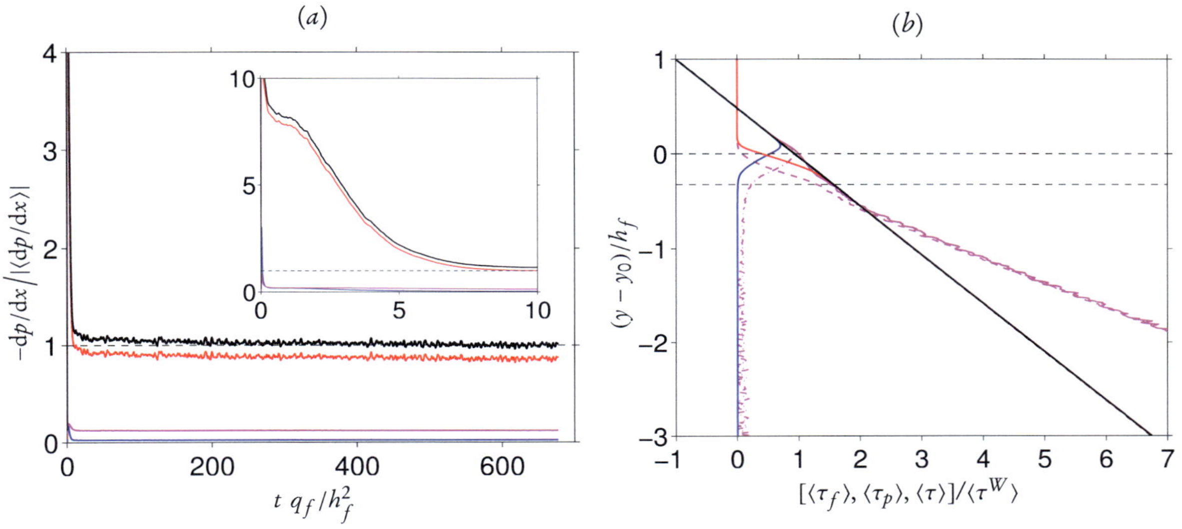

Figure 4.12: (*a*) Time evolution of the imposed pressure gradient (—) which maintains a constant flow rate q_f. —, contribution from the inter-particle collision forces acting on the stationary particles at bottom; —, contribution from the hydrodynamic forces acting on the stationary; —, contribution from the shear stress at the top smooth wall. The mean pressure gradient computed from the steady-state interval is used for normalization. (*b*) Wall-normal profile of the mean fluid shear stress $\langle\tau_f\rangle$ (—), the mean particle shear stress $\langle\tau_p\rangle$ (—), and the mean total shear stress $\langle\tau\rangle$ (—) which perfectly coincides with the line given by $\langle\tau\rangle = \langle\tau^W\rangle - \langle \mathrm{d}p/\mathrm{d}x\rangle(L_y - y)$ (---). The pink colored lines correspond to the fluid and particle stresses as predicted by the constitutive model of Boyer et al. (2011).

As is shown in figure 4.12(*b*), the linearity of the total shear stress profile and its perfect match with the line $\langle\tau\rangle = \langle\tau^W\rangle - \langle \mathrm{d}p/\mathrm{d}x\rangle(L_y - y)$ is confirmed. The figure also shows that in the clear fluid region, the pressure gradient is entirely balanced by the fluid viscous shear stress. On the contrary, the pressure gradient is balanced by the particle shear stress in the bottom region denoted as region III in figure 4.1. In the mobile layer (region II), both contributed to the balance which depends on the wall-normal location. Note that $\langle f_x\rangle$ is a mean of the local Eulerian force density. As is shown in Uhlmann (2008), $\langle f_x\rangle$ is related to the Eulerian average (in wall-parallel bins) of the Lagrangian hydrodynamic force which individual particles experience (integrated over the particle surface and associated with their center location) as:

$$\langle f_x\rangle \approx \langle F_x^H\rangle\langle\phi_s\rangle/V_p \ . \tag{4.22}$$

The error associated with this approximation is not negligible in regions where there is strong gradient in the number of samples per bin and/or strong gradient in $\langle f_x\rangle$.

4.2.5 Comparison with a suspension rheology model

Boyer et al. (2011) describe the effective coefficient of friction as well as the solid volume fraction of a submerged sediment bed, which is confined by a pressure P^p and subjected to a mean constant shear rate $\dot\gamma_p$, as a function of a dimensionless viscous number

$$I_v = \frac{\rho_f \nu \dot\gamma_p}{P^p} \ . \tag{4.23}$$

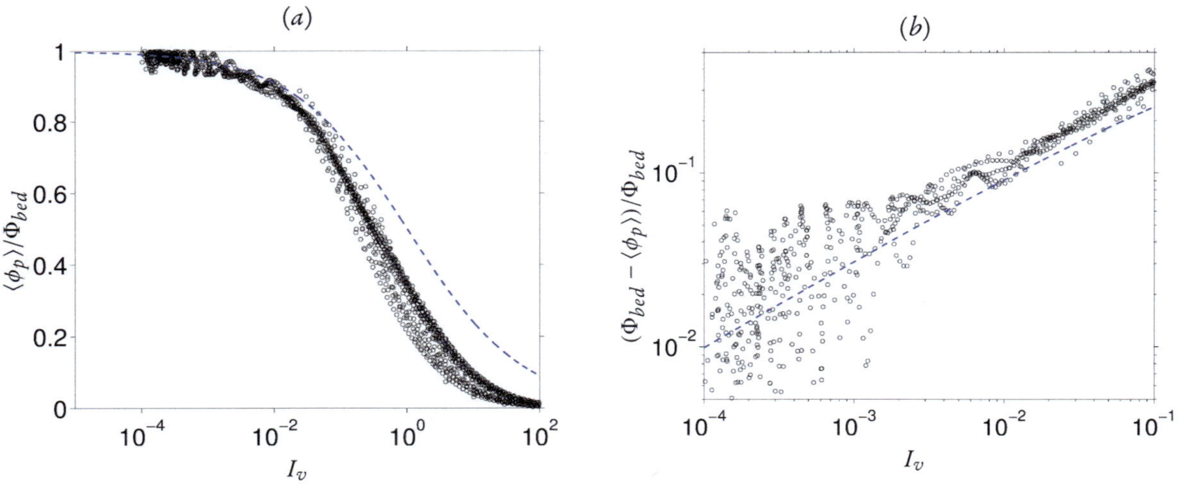

Figure 4.13: (*a*) Mean solid volume fraction of the sediment $\langle\phi_p\rangle$, normalized by the value in the stationary sediment bed Φ_{bed}(cf. equation (4.7)), as a function of the viscous number I_v. The dashed line corresponds to the model proposed by Boyer et al. (2011). (*b*) Asymptotic behavior of $\Phi_{bed} - \langle\phi_p\rangle$ at small values of I_v.

The effective coefficient of friction is defined as

$$\mu_{eff} = \frac{\tau}{P^p} \tag{4.24}$$

where τ is the applied shear stress. For all the experiments they performed, plots of the solid volume fraction $\langle\phi_p\rangle$ and the effective coefficient of friction μ_{eff} versus I_v collapse into respective master curves from which they propose a constitutive law for the evolution of μ_{eff} as:

$$\mu_{eff}(I_v) = \underbrace{\mu_1 + \frac{\mu_2 - \mu_1}{1 + I_0/I_v}}_{\text{contact}} + \underbrace{I_v + \frac{5}{2}\phi_m I_v^{1/2}}_{\text{hydrodynamic}} \tag{4.25}$$

where μ_1, μ_2 and I_0 are constants which depend on material properties and ϕ_m is the maximum solid volume fraction. From their experiments, they determined that $\phi_m = 0.585$ (maximum packing of a homogeneously sheared dense suspension at vanishing I_v), $\mu_1 = 0.32$, $\mu_2 = 0.7$ and $I_0 = 0.005$. The constitutive law (4.25) incorporates contributions from inter-particle contact as well as hydrodynamic stress. Moreover, Boyer et al. (2011) propose a function for the evolution of the solid volume fraction as:

$$\langle\phi_p\rangle(I_v) = \frac{\phi_m}{1 + \sqrt{I_v}} \tag{4.26}$$

which recovers $\phi_m - \langle\phi_p\rangle \propto \sqrt{I_v}$ at small I_v. In the following, we compare the model with the DNS data.

In our numerical configurations, $\phi_m = \Phi_{bed}$. Moreover, unlike in the experimental setup of Boyer et al. (2011), P^p, τ, $\dot{\gamma}_p$ are functions of the wall-normal location. Thus for each simulation case, we have wall-normal profile of I_v, $\langle\phi_p\rangle$ and μ_{eff}. The confining pressure P^p is proportional to the apparent weight of the sediment bed (Aussillous et al., 2013). Thus, for a wall-normal varying solid volume fraction, P^p is given as follows:

$$P^p(y) = (\rho_p - \rho_f)|\mathbf{g}| \int_y^{L_y} \langle\phi_p\rangle(y)\mathrm{d}y\,. \tag{4.27}$$

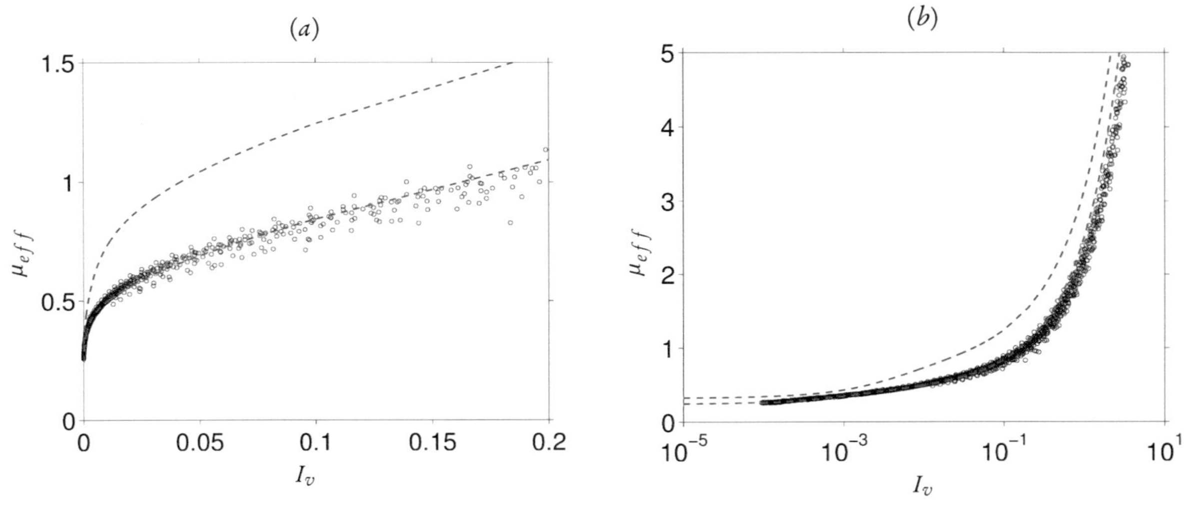

Figure 4.14: Effective friction coefficient of the sediment bed as a function of the viscous number. The blue dashed line corresponds to the constitutive model of Boyer et al. (2011) adopting their parameter values ($\mu_1 = 0.32$, $\mu_1 = 0.70$, $I_0 = 0.005$). The red dashed line is the same model but adopting different set of parameter values ($\mu_1 = 0.24$, $\mu_1 = 0.39$, $I_0 = 0.001$).

Additionally, inside the sediment bed, $\langle u_f \rangle \approx \langle u_p \rangle$, which means $\dot{\gamma}_p(y) \approx \dot{\gamma}_f(y)$. Figure 4.13 shows the variation of the normalized mean solid volume fraction as a function of I_v for the data taken in the mobile sediment bed region of all the simulated cases. The relevance of the viscous number is highlighted by the collapse of the data (exhibiting a slight scatter). Note that the maximum value of the solid volume fraction inside the stationary sediment bed of each case is used for normalization (cf. table 4.1). Due to the conditions of our simulations, Φ_{bed} is somewhat smaller in value than $\phi_m = 0.585$ as explained in section 4.1.3. At small values of I_v, the DNS data marginally matches the model by Boyer et al. (2011). In particular, the evolution $\Phi_{bed} - \langle \phi_p \rangle \propto \sqrt{I_v}$ is fairly recovered as seen in figure 4.13(*b*). At large values of I_v however, the DNS data substantially deviates from the model.

Figure 4.14 shows the variation of the effective coefficient of friction μ_{eff} as a function of the viscous number again for the data of the mobile layer of all simulated cases. This quantity also fairly collapses into a single curve for a wide range of I_v. Nevertheless, the DNS data is seen not to match the model proposed by (Boyer et al., 2011) when using the parameter values consistent with their experimental data ($\mu_1 = 0.32$, $\mu_1 = 0.70$, $I_0 = 0.005$). On the other hand, when using $\mu_1 = 0.24$, $\mu_1 = 0.39$ and $I_0 = 0.001$, the model agrees well with the DNS data. Note that these new set of parameter values are identical to those adopted by Aussillous et al. (2013) except for the value of I_0, for which Aussillous et al. (2013) adopted $I_0 = 0.01$.

For comparison, the wall-normal profiles of the hydrodynamic and contact contribution to the total shear stress balance corresponding to case BL21, as predicted by the constitutive model (4.25), are plotted in figure 4.12(*b*) in pink colored lines. In the mobile layer region, the driving force of the pressure gradient is larger than the resisting friction force stemming from the inter-particle contact. Thus the bed is in motion. The balance between the driving force and total shear resistance from hydrodynamic contribution (considering Einstein viscosity) and contact contribution, in the mobile sediment bed region, is fairly captured by the model. The model also satisfactorily predicts that inside the stationary bed, the inter-particle friction force ($\approx \mu_1 P^p$) is larger in magnitude than the driving force of the pressure gradient. Thus a higher pressure gradient would be required to mobilize particles in this region.

4.3 Summary and conclusion

In this chapter, we have presented results of a direct numerical simulation performed of horizontal channel flow over a thick bed of mobile sediment particles by employing the DNS–DEM coupled numerical algorithm. We have carried out a series of fully-resolved simulations of bedload transport by laminar flow in a plane channel configuration similar to the experiment of Aussillous et al. (2013). Although the Reynolds number in the simulations is two orders of magnitude larger than in the experiment, the range of values of the principal control parameters of the problem (either the Shields number or the non-dimensional flow rate) overlaps significantly between both approaches, allowing for a direct comparison. Our DNS-DEM method was found to provide results which are fully consistent with the available data from the reference experiment. The present simulations yield a cubic variation of the particle flow rate (normalized by the square of the Galileo number times viscosity) with the Shields number, once the threshold value of the Shields number is exceeded. The thickness of the mobile particle layer (normalized with the particle diameter) obtained from our simulations varies roughly with the square of the Shields number; when normalized with the height of the fluid layer, it follows even more closely a quadratic variation with the normalized fluid flow rate. Previous studies using two-fluid models have predicted a linear dependency of the mobile layer thickness on both parameters (Mouilleron et al., 2009; Ouriemi et al., 2009b). Unfortunately, the data for this quantity extracted from the experiment of Aussillous et al. (2013) features a certain amount of scatter, and, therefore, does not serve to clearly distinguish between the two power laws.

Based upon the DNS-DEM results we were also able to show that the fluid mean velocity exhibits a parabolic profile in most part of the clear fluid region. Near the fluid-bed interface, the deviation from a parabolic flow profile begins to become significant. We have defined appropriate velocity and length scales which characterize the streamwise fluid (and particle) velocity profile in the mobile granular layer region. The value of the mean fluid velocity at the fluid-bed interface U_f^0, which is the maximum velocity in the region, is chosen as a velocity scale. For the length scale, the wall-normal offset from the fluid-bed interface to the location where the mean fluid velocity reduces to $0.5U_f^0$, which is termed the characteristic half thickness of the mobile bed $\tilde{h}$, was defined. In the mobile granular layer, the fluid velocity profiles of all simulations cases, when normalized with above mentioned scales, were observed to collapse into a master curve exhibiting only small scatter. It turns out that the fluid mean velocity at the interface exhibit a quadratic relationship with the shields number (or the non-dimensional fluid flow rate) while the characteristic half thickness exhibits a linear variation. The results obtained from our simulations showed satisfactory agreement with those of the experiment by Aussillous et al. (2013), although the latter exhibited substantial scatter with respect to these quantities. Finally, an empirical exponential curve, fully defined by U_f^0 and $\tilde{h}$, is proposed for the particle velocity profiles.

Based on the analysis of hydrodynamic and collision forces acting upon the particles as well as the fluid viscous stress, the exchange of momentum between the phases was addressed. Inside the stationary sediment bed, the driving pressure gradient is entirely balanced by the inter-particle collision forces, while in the clear fluid region, the fluid viscous stress is responsible for balancing the pressure gradient. In the mobile granular layer on the other hand, both fluid and particle stresses contribute to balance the pressure gradient. The stress balance was finally compared to the suspension rheology model proposed by Boyer et al. (2011) in which the effective coefficient of friction μ_{eff} as well as the solid volume fraction $\langle\phi_p\rangle$ of a submerged sediment bed were described as a function of a dimensionless viscous number I_v. The relevance of the viscous number is confirmed with the DNS data which exhibited a collapse of $\mu_{eff}(I_v)$ and $\langle\phi_p\rangle(I_v)$ into respective master curves.

Chapter 5

Pattern formation over subaqueous sediment bed in channel flow

In this chapter, we study the development of subaqueous bedforms over a sediment bed in a statistically unidirectional closed and open channel flow both in the laminar and turbulent regimes. The objective of this study is, first of all, to capture the initiation and evolution of subaqueous bedform patterns by performing interface-resolved direct numerical simulations (DNS) and to provide a complete information of the flow field as well as individual particle-related quantities. Furthermore, based on the analysis of the generated DNS data, this study aims to contribute to the elucidation of the physical mechanisms involved in the formation of sedimentary patterns driven by (turbulent) shear flow. More specifically, in this chapter, an attempt is made to answer the following key questions: For a given set of parameter points,

- Does an initially flat sediment remain stable or become unstable?
- At what initial wavelength do ripples/dunes form?
- How do bedforms evolve during the subsequent phases of the initial instability?
- How does a given computational domain size influence the formation of patterns?
- What is the minimum box size which accommodates an unstable bed?

To this end, we have performed ten independent simulations, covering four sets of parameter values, three sets in the laminar regime (one simulation each) and one set in the turbulent regime (a total of seven simulations). The sediment bed is represented by a large number of freely moving spherical particles; total number of particles per simulation ranges from $O(10^4)$ to $O(10^6)$. All the relevant length and time scales of the flow, even the near-field around the individual particles, are resolved. The parameter points are chosen such that all are above the treshold for particle motion. When compared to the regime classification based on the pipe flow experiment by (Ouriemi et al., 2009a), one parameter point falls very close to the predicted treshold for pattern formation, two sets fall within the laminar 'small dune' regime and one lies within the turbulent 'vortex dune' regime. Figure 5.1 shows the four parameter points considered in the present study in comparison with the experimental data obtained from the pipe flow experiment of Ouriemi et al. (2009a) and channel flow experiment of Langlois and Valance (2007).

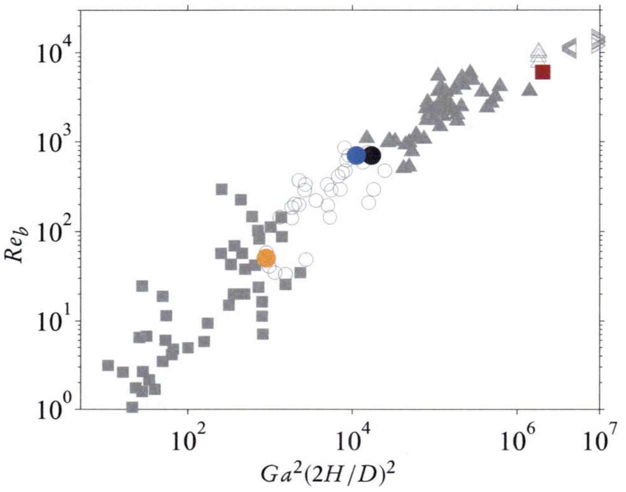

Figure 5.1: Different regimes of sediment bed patterns obtained in the pipe flow experiment of Ouriemi et al. (2009a), shown in the parameter plane (Re_b,$Ga^2(2H/D)^2$): 'flat bed in motion' (■); 'small dunes' (○); 'vortex dunes' (▲). For the pipe flow data the Reynolds number Re_{pipe} based upon the pipe diameter d_{pipe} and the bulk velocity q_f/d_{pipe} is used. The data points in the turbulent channel flow experiment of Langlois and Valance (2007) are indicated by: $D = 100\mu m$ (△); $D = 250\mu m$ (◁); $D = 500\mu m$ (▷). The following symbols refer to the present simulations: ●, L1; ●, L2; ●, L3; ●, T1.

Let us take this opportunity to mention, that part of the results presented in this chapter has been recently published (Kidanemariam and Uhlmann, 2014a). That publication was selected for extended review and was featured in the 'Focus on Fluids' article by Colombini (2014) highlighting the novelty and impact of our work.

5.1 Computational setup

5.1.1 Flow configuration and parameter values

As shown in figure 5.2 a Cartesian coordinate system is adopted such that x, y, and z are the streamwise, wall-normal and spanwise directions respectively. Mean flow and gravity are directed in the positive x and the negative y directions respectively. The computational domain is periodic in the streamwise and spanwise directions. The no-slip/free-slip condition is imposed at the top boundary (depending on the case) while the no-slip condition is imposed at the bottom walls.

In all the cases, the channel is driven by a horizontal mean pressure gradient which is adjusted at each time step in order to impose a constant flow rate q_f. This results in a shearing flow of fluid height h_f over a mobile bed of height h_b. The bulk Reynolds number of the flow is defined based on the total height of the fluid as

$$Re_b = \frac{2u_bH}{\nu}, \tag{5.1}$$

where $u_b \equiv q_f/\langle h_f\rangle_{zxt}$ is the bulk velocity, H is the channel half-height and ν is the kinematic viscosity. Note that $H = \langle h_f\rangle_{zxt}$ in the open channel flow cases, where as $H = 0.5\langle h_f\rangle_{zxt}$ in the closed channel cases. $\langle h_b\rangle_{zxt}$ is the space and time averaged total fluid height. The averaging operator $\langle\cdot\rangle_{zxt}$ is precisely defined in appendix D and the extraction of the fluid-bed interface is discussed in section 5.2.1 in more detail. The

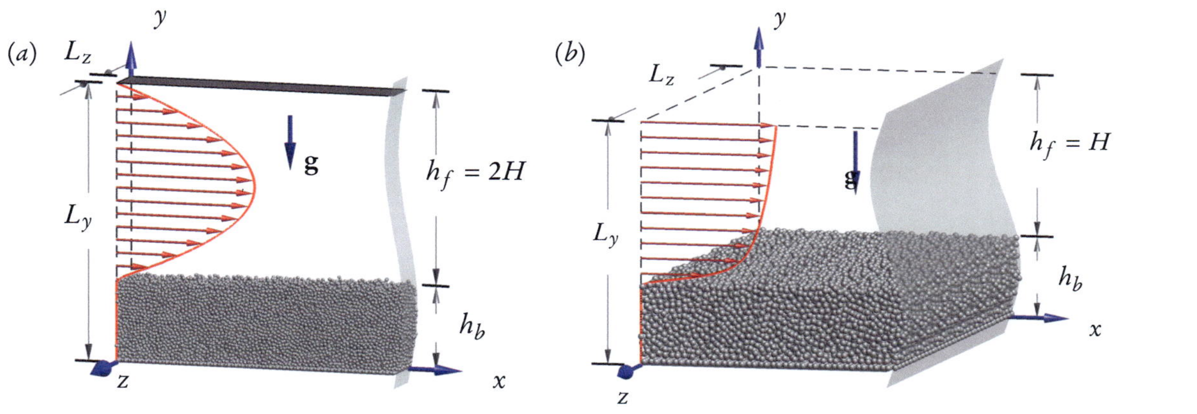

Figure 5.2: Schematic diagrams showing the configuration of the simulations. (*a*) cases L1, L2 and L3. (*b*) case T1. The computational domains are periodic along the x- and z-directions.

friction Reynolds number Re_τ is similarly defined based on the friction velocity u_τ and H, where u_τ is the friction velocity which was defined in equation (2.19). The determination of the friction velocity is based on the extrapolation of the total shear stress of the fluid phase to the time averaged wall-normal location of the fluid-bed interface $\langle h_b \rangle_{zxt}$ (cf. chapter 6 for more details). Other relevant parameters include the particle-to-fluid density ratio ρ_p/ρ_f, the Galileo number Ga, which was defined in equation 2.22, the Shields number Θ, defined in equation (2.21), and a length scale ratio H/D. A relation among the non-dimensional numbers can be formed as

$$\frac{1}{\sqrt{\Theta}} = Re_\tau \frac{1}{Ga} \frac{H}{D} \,. \tag{5.2}$$

To be able to set the bed in motion, the value of the imposed shear stress has to be above the critical value for incipient motion which is determined by the value of the Shields number (cf. section 2.2.1). As was discussed in detail in section 2.2.1, in laminar flows, $\Theta_c = 0.12$. In turbulent flows the critical value is believed to be approximately 0.05 for $D^+ = O(1)$ (Wong and Parker, 2006; Franklin and Charru, 2011). In all the simulations, the density ratio is fixed at a value of $\rho_p/\rho_f = 2.5$ which is approximately equal to the density ratio between glass beads and water. Re_τ is a derived quantity and it is estimated from the imposed mean pressure gradient. In order to accommodate the anticipated bed patterns, the thickness of the fluid height as well as that of the erodible bed is chosen to be sufficiently large when compared to the particle diameter setting the criteria for fixing H/D. This results in adopting a very large number of spherical particles. Note that, in order to create a rough bottom, the first layer of particles are set to be stationary. The arrangement of the fixed layer is similar to that of the sediment erosion cases. Finally, the value of the Galileo number is chosen, by adjusting the value of acceleration due to gravity $|\mathbf{g}|$, such that the value of the Shields number is above the critical value Θ_c(using relation 5.2). Table 5.1 summarizes the values of the relevant physical and numerical parameters adopted for the total of ten simulations performed in this study.

5.1.2 Initiation of the simulations

The simulations were initiated in the following procedure. First a flat granular bed, which is composed of individual mobile spherical particles, is prepared. For this purpose, a discrete element simulation of a large number of spheres was performed considering dry granular flow where the motion of the particles is solely

regime	Case	Re_b	Re_τ	ρ_p/ρ_f	Ga	D^+	H/D	Θ	Θ_{Pois}	channel
	L1	50	8.5	2.5	0.60	0.34	24.92	0.32	0.34	open
Laminar	L2	700	32.3	2.5	2.42	1.20	26.92	0.25	0.25	closed
	L3	700	32.0	2.5	1.97	1.20	26.71	0.37	0.38	closed
	T1	6022	301.7	2.5	28.37	12.06	25.02	0.18	–	open
	$T1^{2H}$	6022	245.8	2.5	28.37	9.76	25.18	0.12	–	open
	$T1^{3H}$	6022	245.7	2.5	28.37	9.68	25.38	0.12	–	open
Turbulent	$T1^{4H}$	6022	252.9	2.5	28.37	10.07	25.12	0.13	–	open
	$T1^{6H}$	6022	263.8	2.5	28.37	10.36	25.45	0.13	–	open
	$T1^{7H}$	6022	260.4	2.5	28.37	10.26	25.38	0.13	–	open
	$T1^{48H}$	6022	260.0	2.5	28.37	10.33	25.16	0.13	–	open

Case	$[L_x \times L_y \times L_z]/D$	Δx^+	$D/\Delta x$	N_p	$T_{obs}u_b/H$
L1	307.2 × 38.4 × 12.8	0.03	10	44471	457
L2	307.2 × 76.8 × 12.8	0.12	10	79073	959
L3	307.2 × 76.8 × 12.8	0.12	10	79073	941
T1	307.2 × 38.4 × 76.8	1.17	10	263 412	1259
$T1^{2H}$	51.2 × 38.4 × 76.8	0.98	10	43 730	259
$T1^{3H}$	76.8 × 38.4 × 76.8	0.97	10	65 359	264
$T1^{4H}$	102.4 × 38.4 × 76.8	1.00	10	86 645	288
$T1^{6H}$	153.6 × 38.4 × 76.8	1.04	10	127 070	274
$T1^{7H}$	179.2 × 38.4 × 76.8	0.96	10	150 521	296
$T1^{48H}$	1228.8 × 38.4 × 76.8	0.97	10	1 053 648	234

Table 5.1: Physical and numerical parameters of the simulations. Re_b and Re_τ are the bulk and friction Reynolds numbers respectively. ρ_p/ρ_f is the particle-to-fluid density ratio. Ga and Θ are the Galileo number and and the Shields number respectively. Θ_{Pois} the Shields number computed under the assumption of a parabolic Poiseuille profile. L_i is the domain length adopted in the i-th direction. In all the cases, the spherical particles have a diameter $D = 10/256$ and a uniform grid spacing $\Delta x = \Delta y = \Delta z = 1/256$ is adopted yielding the above listed resolutions. N_p is the number of spherical particles adopted. T_{obs} is the observation time.

due to forces arising from particle-/wall-particle collisions and gravity. The spheres, which were initally randomly distributed in space, settle towards a bottom flat wall and form a pseudo-randomly packed bed of desired initial bed thickness h_b above the bottom wall.

For the laminar cases, a Poiseuille flow field $\mathbf{u}_0(\mathbf{x})$ with the prescribed flowrate is imposed such that the streamwise component of $\mathbf{u}_0$ has a parabolic profile in the wall-normal interval $y_0 < y < L_y$ and $\mathbf{u}_0 = 0$ in the $0 < y < y_0$. After the simulation starts, the sediment bed immediately adopts to the flow and erosion occurs.

The initiation of the simulation for the turbulent cases was performed in several stages in order to ensure that the flow field is fully turbulent before imposing it on the mobile bed. To this end, a flow field $\mathbf{u}_0(\mathbf{x})$ is prepared which is a superposition of a base laminar Poiseuille flow (with the prescribed flowrate) and a purturbation field $\mathbf{u}_0'(\mathbf{x})$ with zero mean. Similar as in the laminar cases, $\mathbf{u}_0 = 0$ for $y < y_0$ and has parabolic mean for $y > y_0$. In order to facilitate the transition period towards the desired turbulent state, in the first phase of the simulation, the particles were fixed in space while the flow evolves on a computational grid which was coarser by a factor of two than the current grid resolution. Moreover, the purturbation field $\mathbf{u}_0'$ was obtained by interpolation from an already available turbulent flow field. When the flow field was fully developed in the coarse grid, it was linearly interpolated onto the current finer grid. The simulation was then continued while still keeping the particles fixed in space. Particles were first allowed to freely move after the flow field has reached a fully developed turbulent state.

5.2 Spatial and temporal evolution of the sediment bed

Visualizations of a sequence of particle snapshots as shown in figure 5.3 show, after the particles are released and after the initial dilation of the sediment bed, the top layer of the mobile particles start eroding and adapt to the shear imposed by the flow in a relatively short time. Sediment bed erosion-deposition process then takes place where particles are observed to be in motion by rolling, sliding and to a lesser degree saltating within and in the vicinity of the mobile sediment layer. In the turbulent cases, few particles occasionally get ejected into the outer flow and are transported as suspended load. The continued process of erosion from certain regions and deposition at other downstream regions, introduces small purturbations to the initially flat bed. In most of the cases, this leads to the organization and formation of patterns. The process of emergence of the patterns in the turbulent regime seems to differ when compared to that in the laminar regime.

In the laminar cases, during the initial stages of particle erosion, the spatial distribution of moving particles is almost uniform and the formation of the observed patterns happens in a very gradual process. On the other hand, in the turbulent regime, particles at the top of the sediment bed are observed to immediately get eroded in an inhomogeneous manner, with more erosion occuring in certain streamwise elongated regions. Most of the eroded particles are observed to get deposited at a certain downstream distance, in certain prefered regions of the bed. As a result of this continued preferntial erosion-deposition process, the sediment bed initially exhibits streamwise-aligned alternating troughs and ridges. At certain streamwise locations, the ridges are observed to grow in spanwise dimension which tend to disconnect the troughs and form spanwise-oriented undulations in the bed. These undulations are observed to trigger and facilitate the formation of the patterns at a certain preferred initial wavelength. This difference in the formation process of the patterns between the laminar regime and turbulent regime could be explained by the existence of a the near-wall coherent structures in the turbulent case (Kidanemariam et al., 2013). A movie available online at http://dx.doi.org/10.1017/jfm.2014.284 nicely shows the evolution of the sediment bed.

5.2.1 Extraction of the fluid-bed interface

The fluid-bed interface and thus the fluid height h_f as well as the sediment bed height h_b, are determined in the following manner. First, a wall-perpendicular x - y plane $\mathcal{P}$ located at a z-coordinate $z_{\mathcal{P}}$ is defined such that any point in the plane has a coordinate vector $\mathbf{x}_{\mathcal{P}} = (x,y,z_{\mathcal{P}})^T$. The intersection region of the particles with the plane $\mathcal{P}$ at time t is given by the quantity $\phi_p(\mathbf{x}_{\mathcal{P}},t)$, where ϕ_p is the solid-phase indicator function defined in appendix D. In order to obtain a statistically two-dimensional instantaneous fluid-bed interface, the particle indicator function is averaged in the spanwise direction to yield $\langle\phi_p\rangle_z(x,y,t)$ as shown in figure 5.4(a). Precise definition of the averaging operator in the ith direction $\langle\cdot\rangle_i$ can be found in appendix D . $\langle\phi_p\rangle_z$ is a

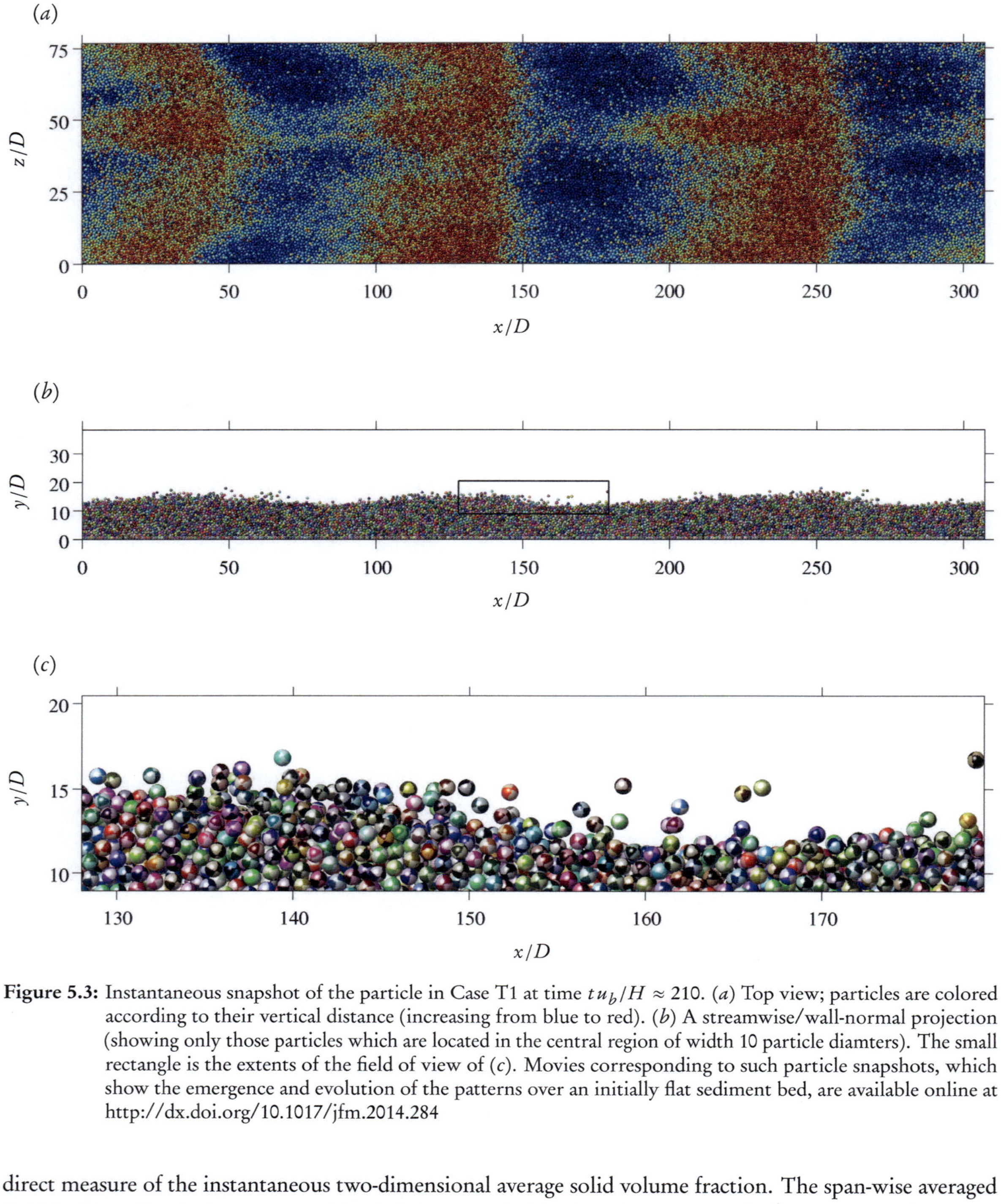

Figure 5.3: Instantaneous snapshot of the particle in Case T1 at time $tu_b/H \approx 210$. (*a*) Top view; particles are colored according to their vertical distance (increasing from blue to red). (*b*) A streamwise/wall-normal projection (showing only those particles which are located in the central region of width 10 particle diamters). The small rectangle is the extents of the field of view of (*c*). Movies corresponding to such particle snapshots, which show the emergence and evolution of the patterns over an initially flat sediment bed, are available online at http://dx.doi.org/10.1017/jfm.2014.284

direct measure of the instantaneous two-dimensional average solid volume fraction. The span-wise averaged fluid-bed interface is then extracted by defining a treshold value chosen as $\langle\phi_p\rangle_z^{tresh} = 0.10$ (see chapter 4). Thus the sediment bed height and the corresponding fluid height are defined as

$$\langle h_b\rangle_z(x,t) = y \mid \langle\phi_p\rangle_z(x,y,t) = \langle\phi_p\rangle_z^{tresh} , \quad (5.3a)$$

$$\langle h_f\rangle_z(x,t) = L_y - h_b(x,t) . \quad (5.3b)$$

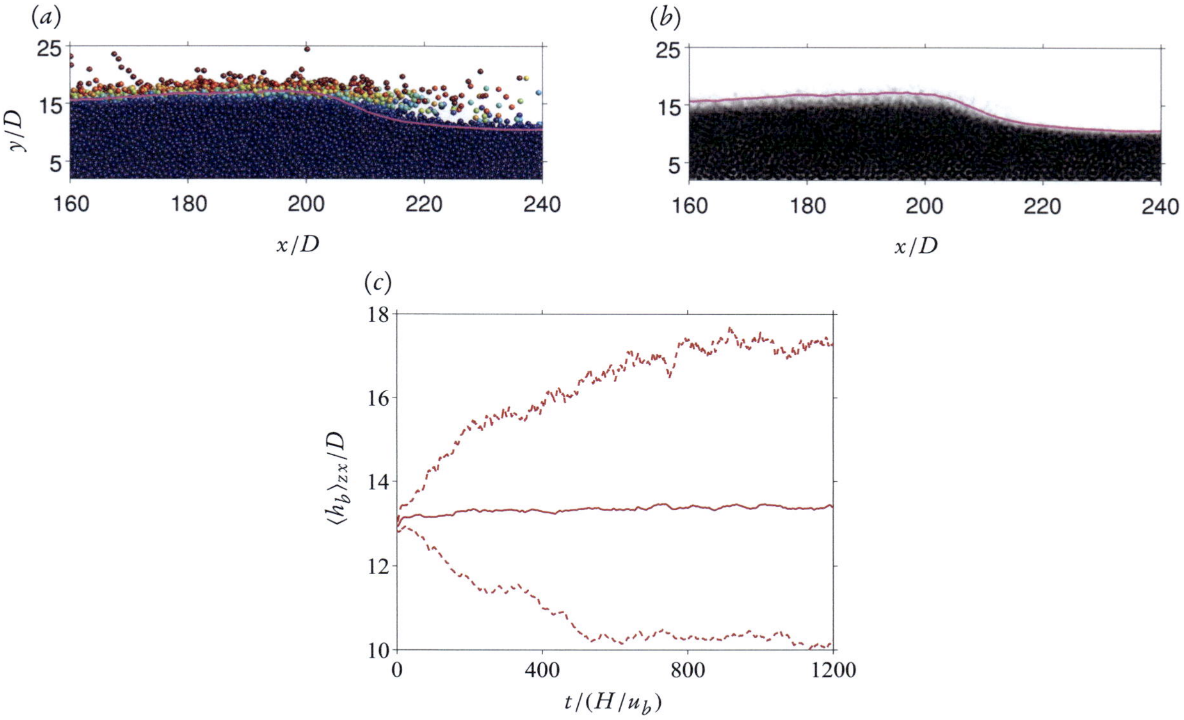

Figure 5.4: (*a*) Close-up showing particles (colored according to their streamwise velocity) for case T1 at an arbitrary time along with its spanwise-averaged solid volume fraction $\langle \phi_p \rangle_z(x,y,t)$. The extracted fluid-bed interface, is shown in pink color. (*b*) Time evolution of the total thickness of the bed. The solid lines correspond to the instantaneous average bed thickness $\langle h_b \rangle_{zx}(t)$ and the dashed lines correspond to the minimum and maximum values of $\langle h_b \rangle_{zx}$ at each time.

Figure 5.4(*b*) shows the time evolution of the instantaneous spatially averaged sediment bed height $\langle h_b \rangle_{zx}(t)$. The evolution of the instantaneous global minimum and maximum values of $\langle h_b \rangle_{zx}$ are also shown in the figure. After a small initial dilation, $\langle h_b \rangle_{zx}$ quickly reaches an almost constant value. However, the increasing deviation with time of the global extremes from $\langle h_b \rangle_{zx}$ indicates that the bed is continuosly evolving.

5.2.2 Space-time variation of the fluid-bed interface

The evolution of the sediment bed and the emergence of patterns as well as the evolution of the two-dimensional shape of the emerged patterns could be infered by scrutinizing the spatial and temporal variation of the sediment bed height fluctuation with respect to the instantaneous average bed height,

$$\langle h_b \rangle'_z(x,t) = \langle h_b \rangle_z(x,t) - \langle h_b \rangle_{zx}(t) \ . \tag{5.4}$$

Figure 5.5 shows the space-time plot of $\langle h_b \rangle'_z$ for the simulations corresponding to the four parameter points. This type of plot provides a qualitative description of the characterstic behaviour of the patterns as a function of space and time. For the duration of the simulated interval, the sediment bed of case L1 seems to evolve featureless without showing any sign of formation of patterns. On the other hand, the spatial and temporal evolution of the two other laminar cases L2 and L3, as well as the turbulent case T1 exhibit the formation of distinct regular undulations of the sediment bed shortly after staring the simulations. The contrast level of

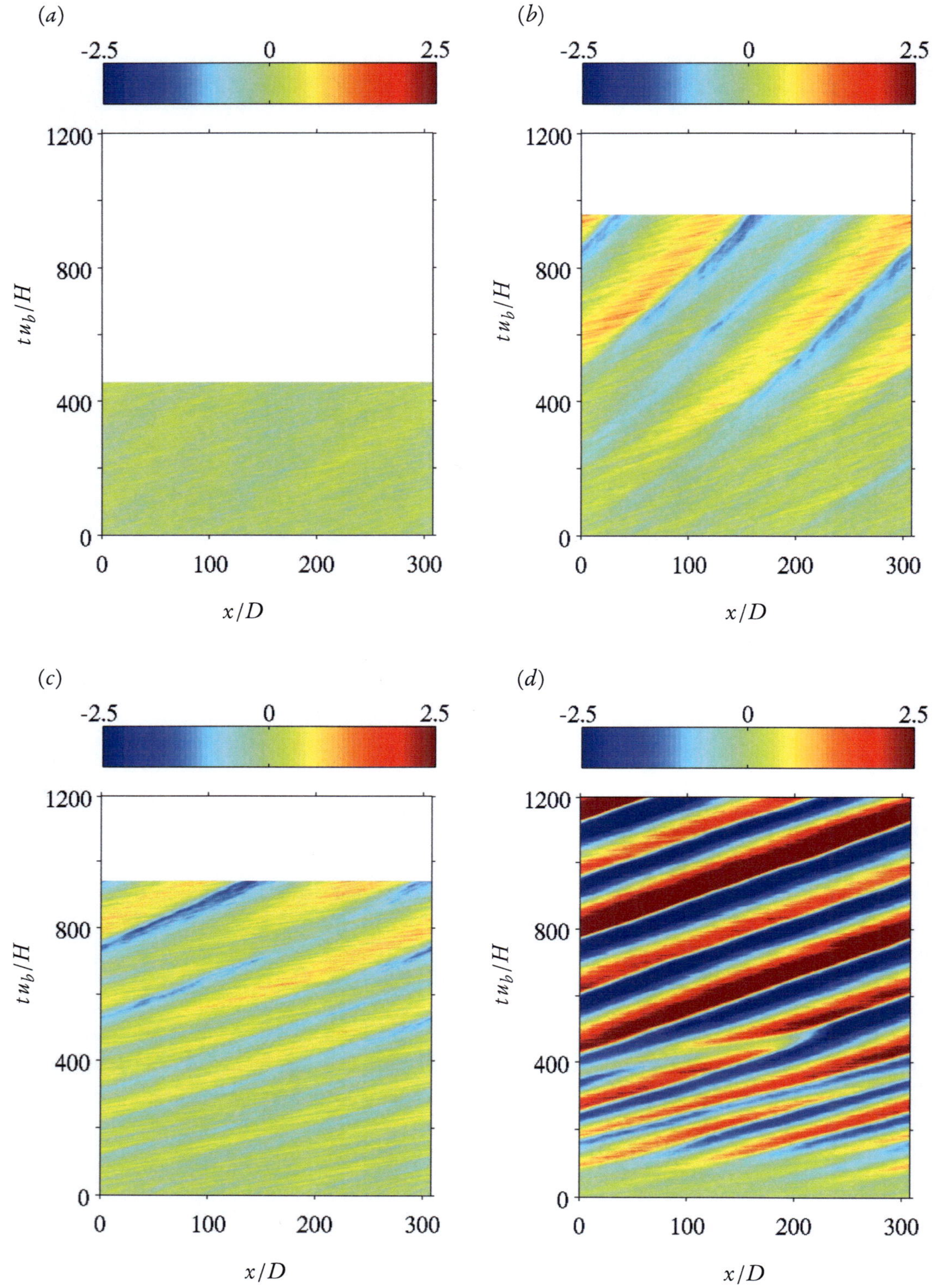

Figure 5.5: Space-time evolution of the fluctuation of the fluid-bed interface $\langle h_b \rangle_z'$, normalized with the particle diameter D. $\langle h_b \rangle_z'$ represents the fluctuation the height of the bed from the instantaneous average bed height $\langle h_b \rangle_{zx}$. (*a*) case L1; (*b*) case L2; (*c*) case L3; (*d*) case T1.

the color coding is a direct measure of the wall-normal height (amplitude) of the formed patterns. Figures 5.5(*b*)-(*d*) qualitatively show that the undulations of the laminar cases L2 and L3 have more or less comparable amplitude which, on the other hand, is clearly smaller than that of the turbulent case T1. Furthermore, the figures also give a qualitative impression of the propagation speed of the formed patterns which are observed to drift in the positive streamwise direction. The bedform features of the turbulent case T1 drift at a larger speed when compared to the laminar cases. A difference in propagation speed between case L2 and L3 is also noticable, patterns in the latter propagating faster than those in the former. The figures also show the process of pattern growth and merger. In the following, we quantify these qualitative observations.

5.2.3 Dimensions of the sediment bed features and their evolution

The most relevant parameters, among others, which describe the geometrical features of two-dimensional bedforms are their streamwise and wall-normal dimensions, namely the wavelength and amplitude of the patterns, as well as their two-dimensional shape. As was noted by Coleman and Nikora (2011), various definitions have been previously adopted to quantify these dimensions. For instance, the average wall-normal distance between local maxima and adjacent local minima of h_b is used as a measure of the amplitude of the patterns in some studies (Ouriemi et al., 2009a). Other studies characterize the amplitude of the patterns by the r.m.s. fluctuation of the bed (Langlois and Valance, 2007) or as a weighted average of the amplitudes of a set of Fourier modes which represent the bed interface (Betat et al., 2002). Similarly, the wavelength of patterns is quantified both geometrically, based on the spacing between alternating troughs or ridges of the bed, and statistically, based on the autocorrelation of the bed height fluctuation Coleman and Nikora (2011).

Mean amplitude of bedforms

In this study, we have adopted the above mentioned methods to quantify the pattern dimensions in order to compare the DNS results with available experimental data. More precisely, the wall-normal dimension of the patterns is characterized by the r.m.s. fluctuation of the sediment bed, defined as

$$\sigma_h(t) = \langle \langle h_b \rangle_z'(x,t) \cdot \langle h_b \rangle_z'(x,t) \rangle_x^{\frac{1}{2}}, \tag{5.5}$$

as well as by the the difference in the wall-normal coordinate of the local maxima and that of adjacent local minima (averaged over the two minima located upstream and downstream of each of the maximum location) denoted as $\langle A_h \rangle_D$. The operator $\langle \cdot \rangle_D$ indicate an average over the number of dunes at a given time. Figure 5.6 shows a sample spanwise averaged instantaneous sediment bed height and the corresponding locations of maxima/minima points for case T1. The temporal evolution of these two quantities, normalized by the particle diameter, for the four sets of parameter points is depicted in figure 5.7(*a*,*b*). As expected, both definitions evolve similarily following a linear relationship. A line, given by $\langle A_h \rangle_D = 2.75\sigma_h$, sufficiently fits the data corresponding to all cases (figure omitted). Thus in the following, the wall-normal height of the bedforms will be addressed by discussing σ_h only. Consistent with the previous discussion, the r.m.s. amplitude of the sediment bed height of case L1 does not grow with time indicating the bed stays stable during the entire observation interval. It exhibits a finite value of $\sigma_h \leq 0.13D$ which corresponds to the random uncorrelated bed fluctuations as a result of the discrete motion of the bed. For instance, Coleman and Nikora (2009) report $\sigma_h/D = 0.17$ for their static flat bed. The bedform amplitude of the other laminar and turbulent cases, on the other hand, is observed to contineously grow as a function of time during the entire simulated duration. The evolution of case L2 and L3 is approximately the same exhibiting a constant growth

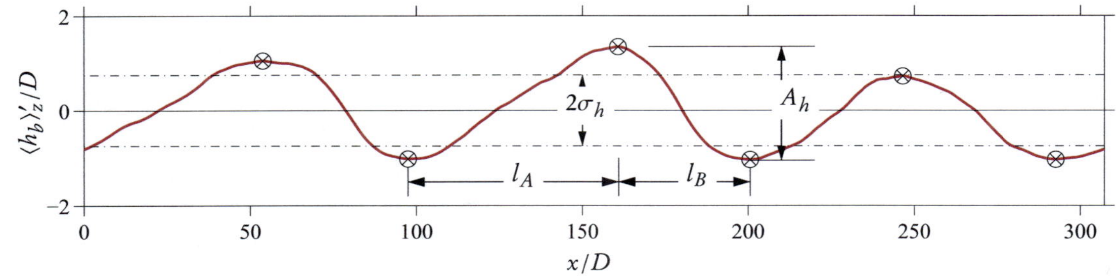

Figure 5.6: Sample streamwise profile of the fluctuation of the spanwise-averaged sediment bed height at time $tu_b/H \approx 140$ for case T1. The markers ⊗ indicate the locations of the local maxima/minima which are the references for the determination of pattern dimensions by the geometrical approach. The dashed lines demarcate the extent of the r.m.s. value for this instant. The streamwise distance from the crest point of a given dune to the trough point located downstream is donoted as l_B. Likewise l_A denotes the same quantity but in the upstream side of the crest. $\langle A_h \rangle_D$ of a given dune denotes the average of the wall-normal distance between the crest and the two local minimum in the downstream and upstream side.

rate of $6.2 \cdot 10^{-4} D$ per unit bulk time and showing no sign of reaching a steady-state regime. In contrast, the amplitude of the turbulent case T1 is observed to grow much faster than that of the laminar cases in the first 200 bulk time units of the simulation exhibiting a growth rate of $6.4 \cdot 10^{-3} Du_b/H$. It then attains a milder growth rate (comparable to that of the laminar cases) for another approximately 250 bulk time units, followed by a sharp increase in amplitude in a relatively short time interval of approximately 100 bulk time units. Finally, until the end of the simulated time, the amplitude grows very slowly, at an average rate of $1.2 \cdot 10^{-4} Du_b/H$, which is even smaller than that of the laminar cases. In fact, for $tu_b/H \gtrsim 800$, σ_h seems to show no substantial growth. As can be seen in figure 5.7(*a*), the time evolution in the final period of case T1 is roughly equivalent to the one determined in a laboratory experimet by Langlois and Valance (2007) at comparable parameter values with $Re_{2H} = 15\,130$, $\Theta = 0.099$, and $H/D = 35$. The order of magnitude difference in the initial growth rates of the amplitude between the turbulent regime ('vortex dunes') and that of the laminar regime ('small dunes') is also reported in the pipe flow experiment of Ouriemi et al. (2009a).

Mean wavelength of bedforms

The streamwise dimension of the bedforms is determined from the instantaneous two-point auto-correlation of the bed height fluctuation as a function of streamwise separation δx, defined as

$$R_h(\delta x,t) = \langle\, \langle h_b \rangle'_z(x,t) \cdot \langle h_b \rangle'_z(x+\delta x,t)\, \rangle_x \ . \tag{5.6}$$

Figure 5.7(*c*) shows the auto-correlation R_h, normalized by σ_h^2, at time $tu_b/H = 250$ for each of the parameter points. It should be noted that, as a result of the adopted boundary condition of the simulations, the relevance of the quantity R_h is limited up to half the streamwise period of the computational domain ($L_x/2$). The spatial correlation, as a function of δx, for case L1 is relatively small when compared to the other cases and does not show any distinct feature confirming the fact that no bedform features are emerging. On the other hand, for cases L2, L3 and T1, the two-point correlation exhibits a distinct damped-oscillation curve featuring, alternating positive and negative values of R_h as a function of δx. The magnitude of the local maxima/minima monotonically dampens with increasing δx from a value of $R_h = 1$ at $\delta x = 0$. Therefore, as is illustrated in figure 5.7(*c*), we can define an average bedform wavelength λ_h as twice the streamwise separation at which the global minimum of R_h occurs viz.

$$\lambda_h(t) = 2x \mid \forall\, \delta x \in [0, L_x/2] : R_h(\delta x,t) \geq R_h(x,t)\,. \tag{5.7}$$

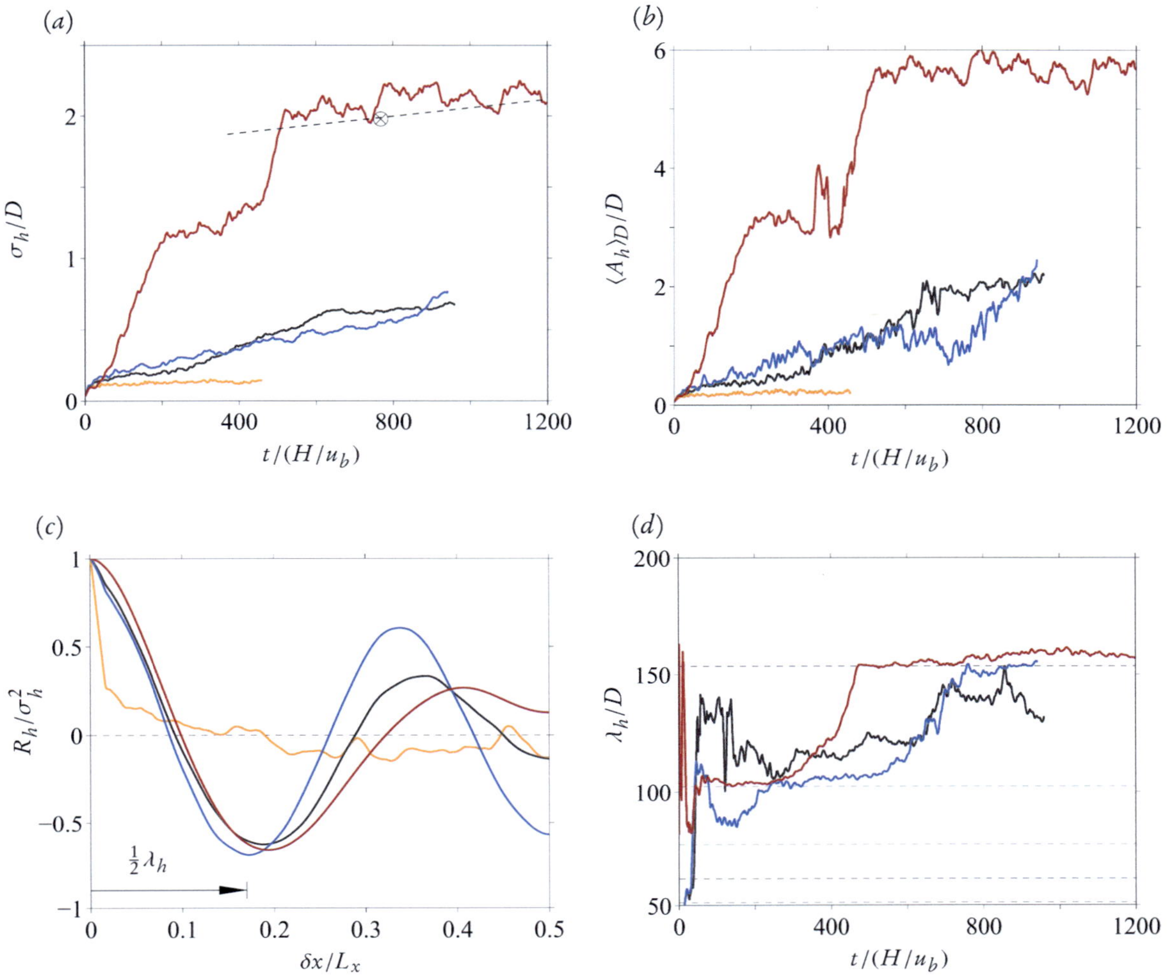

Figure 5.7: (*a*) Time evolution of the r.m.s. sediment bed height normalized by the particle diamter. The dashed line shows the fit obtained by Langlois and Valance (2007, figure 6*a*) for their case with Re_{2H} = 15 130, $\Theta = 0.099$, $H/D = 35$. Note that their first data point starts at time $t \approx 740 u_b/H$ which is indicated by the symbol ⊗. (*b*) Time evolution of the mean wall-normal distance between succesive maxima and minima of the sediment bed height normalized by the particle diameter. (*c*) Two-point auto-correlation fuction as a fuction of the streamwise separation. The definition of the mean wavelength of the bedforms is based on the location of the miminum value of R_h. (*d*) Time evolution of the mean wavelength of the sediment bed height normalized by the particle dimater. The dashed lines indicate the wavelengths of the second to sixth streamwise harmonics. Color coding of all the plots is as follows: —,L1; —,L2; —,L3; —,T1;

The time evolution of λ_h, normalized by the particle diamter, for cases L2, L3 and T1 is shown in figure 5.7(*d*). During the first approximately 40 bulk time units, the evolution of λ_h shows no clear trend showing large oscillations. These oscillations are most probably related to the initial uncorrelated random fluctuations of the sediment bed. Starting from time $t u_b/H \approx 50$, the mean wavelength in case L2 is observed to attain values in the range $\lambda_h/D \approx 110$–150 during the entire simulated interval. The mean wavelength evolves between the second and third harmonics of the computational domain namely $\lambda_i = L_x/2$ and $\lambda_i = L_x/3$ indicating that both modes are growing simultaneously (the wavelengths of the first few resolved harmonics are shown by the dashed lines in figure 5.7(*d*)). On the other hand, the evolution of the wavelength of case L3 seems to be dominated more or less by the harmonic with wavelength $\lambda_i = L_x/3 = 102.4D$ until time $t u_b/H \approx 500$. In the duration between 500 and 750 bulk time units, the wavelength grows at a faster rate and finally settles at $\lambda_i = L_x/2 = 153.6D$ until the end of the simulation. The evolution of the wavelength of the turbulent case T1 also behaves in a similar way as that of case L3 but a relatively shorter time scales. It first settles at $\lambda_i = L_x/3$ and more or less maintains this wavelength until time $tu_b/H \approx 250$. After this time, the average wavelength grows in a relatively short time duration and settles again at a wavelength $\lambda_i = L_x/2$ at time $tu_b/H \approx 450$. It maintains this value until the end of the simulated interval with small deviations of about 5 particle diamters. In turbulent channel flow experiments, Langlois and Valance (2007) have determined values of the initial pattern wavelength $\lambda/D \approx 100 - 150$ which is in good agreement with the present DNS results. Moreover, Ouriemi et al. (2009a) report an initial wavelength of the 'small dunes' in the range $\lambda = 2.5 - 12.6H$ and that of the 'vortex dunes' to be approximately $4H$ (data extracted from figures 3(*c*)and 7(*b*) of their paper). The average pattern wavelengths in cases L2, L3 and T1 measures $3 - 6.5H$ comparing very well with their results.

Bedform morphology

It is well know that bedforms of finite amplitude evolve towards an asymmetric shape even in the absence of flow separation (Best, 2005; Colombini and Stocchino, 2008; Seminara, 2010). For instance, spatio-temporal plots provided by Ouriemi et al. (2009a) have evidenced that the evolution of the bedforms towards their asymmetric shape is a phenomenon of both the laminar 'small dunes' which exhibit no flow separation in their leeside as well as the turbulent 'vortex dunes' which are characterized by flow separation downstream of their crests. The animations of particle motion provided by Kidanemariam and Uhlmann (2014a) as well as the space-time plots depicted in figure 5.5 have qualitatively shown the asymmetric evolution of bed shape in good agreement with what is observed experimentally. In order to give a quantitative measure of the evolution of the two-dimensional shape, two length scale ratios are defined as:

$$AR(t) = \left\langle \frac{A_h}{l_A + l_B} \right\rangle_D(t) , \quad (5.8a)$$

$$LR(t) = \left\langle \frac{l_B}{l_A + l_B} \right\rangle_D(t) . \quad (5.8b)$$

The definition of the quantities A_h, l_A and l_B is given in figure 5.6. The ratios AR and LR give a measure of the aspect ratio and degree of asymmetry of the bedform features. It should be noted that, quantifying the skewness of the dune shapes by a symmetry index as provided in equation (5.8) has been done previously (see e.g. Blondeaux et al., 2015). Figure 5.8 shows the time evolution of these two quantities for the three cases L2, L3 and T1. The aspect ratio of the laminar cases exhibit the same trend, AR varying approximately linearly for the entire simulated interval. A small change in the growth rate is observed at time $tu_b/H \approx 600$. They attain a value $AR \approx 1.5\%$ at the end of the simulation time. Case T1 on the other hand shows two plateau-like regions which are associated to the above mentioned evolution of the wavelength and amplitude of the patterns. The aspect ratio attains a some what constant value of $AR \approx 3.7\%$ after time $tu_b/H \approx 500$ until the end of the simulated interval. The aspect ratio of fully developed subaqueous sediment ripples has been reported to be approximately 1/15 (Charru et al., 2013). This value is almost twice that of the turbulent case of the present simulations. This could be an indication that the ripples have not yet fully developed.

Now let us turn to the asymmetry of the patterns. In case of a symmetric bedform shape, the length scale ratio LR is expected to be equal to 0.5. Thus, the deviation of LR from its symmetric value as a function of time gives a quantitative measure of the evolution of the bedforms towards their asymmetric shape. Figure 5.8(*b*) shows, that during the initial bed instability, the bedforms exhibit an approximately symmertic shape.

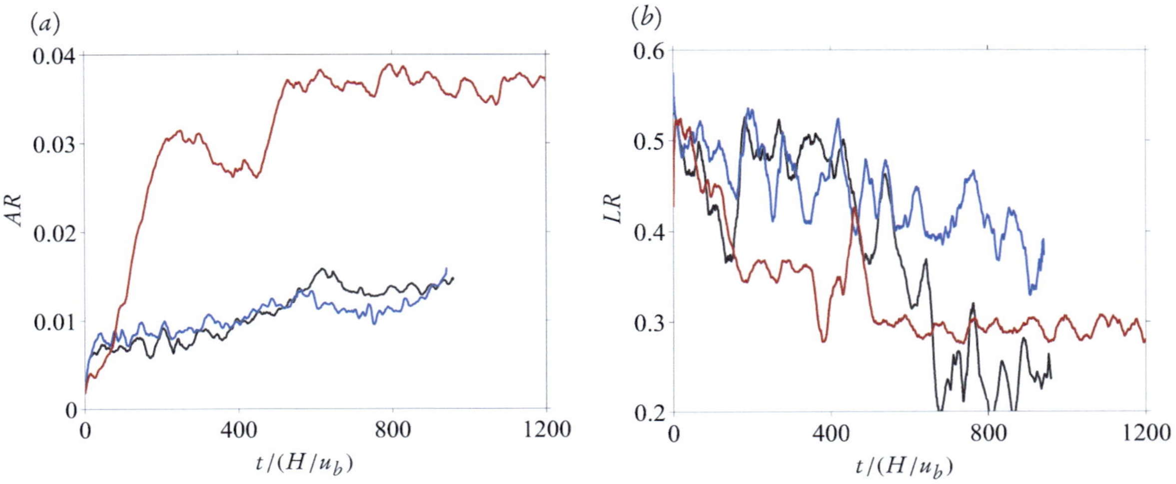

Figure 5.8: (*a*) Time evolution of the ratio of of the wall-normal dimension to the streamwise dimension of the bedform features. (*b*) Time evolution of the degree of asymmetry of the two-dimensional shape of the bedforms as defined in (5.8). —,L2; —,L3; —,T1.

Nevertheless, the bedforms become skewed in a relatively short time after their initial formation, and the evolution of the asymmetry seems to vary among the different cases. During the simulated time interval, the variation of LR of case L2 seems to show a relatively larger fluctuations when compared to cases L3 and T1. LR first decreases at a milder rate, then drops at a larger rate during the time interval $tu_b/H \approx 400-700$. It then maintains a relatively constant value of $LR \approx 0.25$ until the end of the simulated time. LR of case L3, on the other hand, seems to gradually decrease from its initial value of $LR \approx 0.5$ almost at a relatively constant rate, also exhibiting a certain degree of oscillation, and finally reaches a value of $LR \approx 0.4$ at the end of the simulated time. The turbulent case T1 shows a relatively stable evolution of LR, decreasing from its initial value relatively quickly during the first approximately 200 bulk time units, and asymptotically attaining a value of $LR \approx 0.3$ at time $tu_b/H \approx 500$. It constantly maintains this value until the end of the simulated time.

5.2.4 Propagation speed of the bedforms

The propagation speed of bedforms is an important quantity of interest in the study of morphodynamics. For instance, the rate at which bedforms propagate is closely related to the rate of sediment transport (Nikora et al., 1997). We devote this section to study the time evolution of this quantity based on our DNS data (corresponding to the dune-featuring simulation cases L2, L3 and T1).

The propagation speed of the patterns can be determined from the space-time correlation function of the fluid-bed interface fluctuation $\langle h_b \rangle'_z$ (see e.g. Nikora et al., 1997; Coleman and Nikora, 2011). Note that a space-time correlation fuction is more suited to a statistically stationary data, whereas the present study focuses mainly on the regime in which the patterns are still developing, and their propagation speed varies with time. In order to determine the time-dependent propagation speed, the entire observation interval is decomposed into smaller time intervals of $\Delta t_d \approx 25H/u_b$. The evolution time scale of the patterns is much larger than Δt_d and thus, within a given interval, the dimensions of the patterns are expected to not vary substantially. Then, a time-varying space-time correlation function is defined as

$$R_{ht}(\delta x,\delta t,t_i) = \langle \langle h_b \rangle'_z(x,t') \cdot \langle h_b \rangle'_z(x+r_x,t'+\delta t) \rangle_{xt} \quad \forall t' \in [t_i - {}^1\!/\!_2\Delta t_d,\, t_i + {}^1\!/\!_2\Delta t_d] \tag{5.9}$$

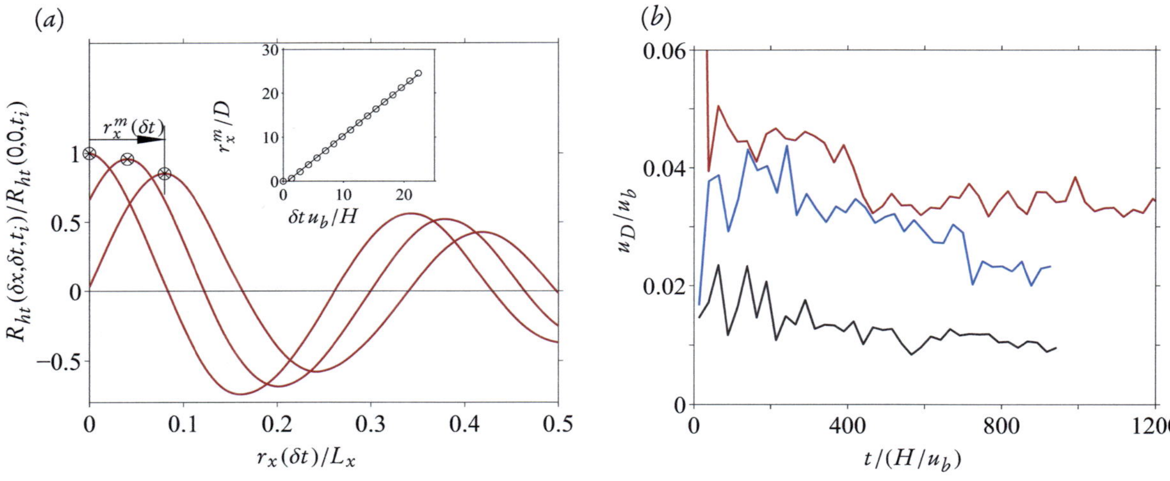

Figure 5.9: (*a*) Sample space-time correlation function plots computed within a given time interval of $\Delta t_d \approx 25H/u_b$. The markers indicate the location of the global maximum of R_{ht} for a given time separation δt. The inset shows the variation of the location of the maximum as a function of δt. (*b*) Time evolution of the propagation speed of the patterns normalized by the bulk velocity. —,L2; —,L3; —,T1;

where t_i is the mean time of the ith interval. Within each interval, the streamwise location of the maximum of R_{ht}, which is given by

$$r_x^m(\delta t,t_i) = x \mid \forall\, \delta x \in [-L_x/2,\, L_x/2] : R_{ht}(\delta x,\delta t,t_i) \leq R_{ht}(x,\delta t,t_i)\,, \tag{5.10}$$

varies approximately linearly with δt (cf. figure 5.9(*a*)), and the slope of the line which is fitted to the data r_x^m versus δt gives the average bedform propagation speed within the interval.

Figure 5.9(*b*) shows the time evolution of the propagation speed of the patterns u_D computed by the above mentioned procedure. The patterns of all the three cases propagate at a speed which is observed to decreases with time. More specifically, the propagation speed of the patterns in the laminar case L2 decreases mildly from an initial value of approximately $0.02u_b$ to a final value of approximately $0.01u_b$, while that of cases L3 decreases from a value of approximately $0.04u_b$ to a final value of $0.022u_b$, with u_D of case L3 roughly twice as large as that of case L2. Similarly, the propagation speed of the turbulent case T1 decreases from an initial value $u_D \approx 0.045u_b$. In the final period of the simulation (the period in which the dune dimensions were observed to show negligible evolution), u_D attains a more or less constant value of approximately $0.034u_b$. This value is consistent with the range of values reported for 'vortex dunes' by Ouriemi et al. (2009a, figure 3*b*). This means that, in relation to the observations made in the previous section, the propagation speed is inversely proportional to the dune dimensions, with a proportionality coefficient which is different for each of the cases. This finding is consistent the literature (see e.g. Betat et al., 2002; Coleman et al., 2003; Ouriemi et al., 2009a).

5.2.5 Spectral description of the sediment bed height

The auto-correlation function of the sediment bed height fluctuation provides integral statistical information such as the time and length scales of the evolving features. On the other hand, a spectral description of $\langle h_b \rangle_z$ provides additional detailed information, for instance on the dispersion relation, which is a description of the

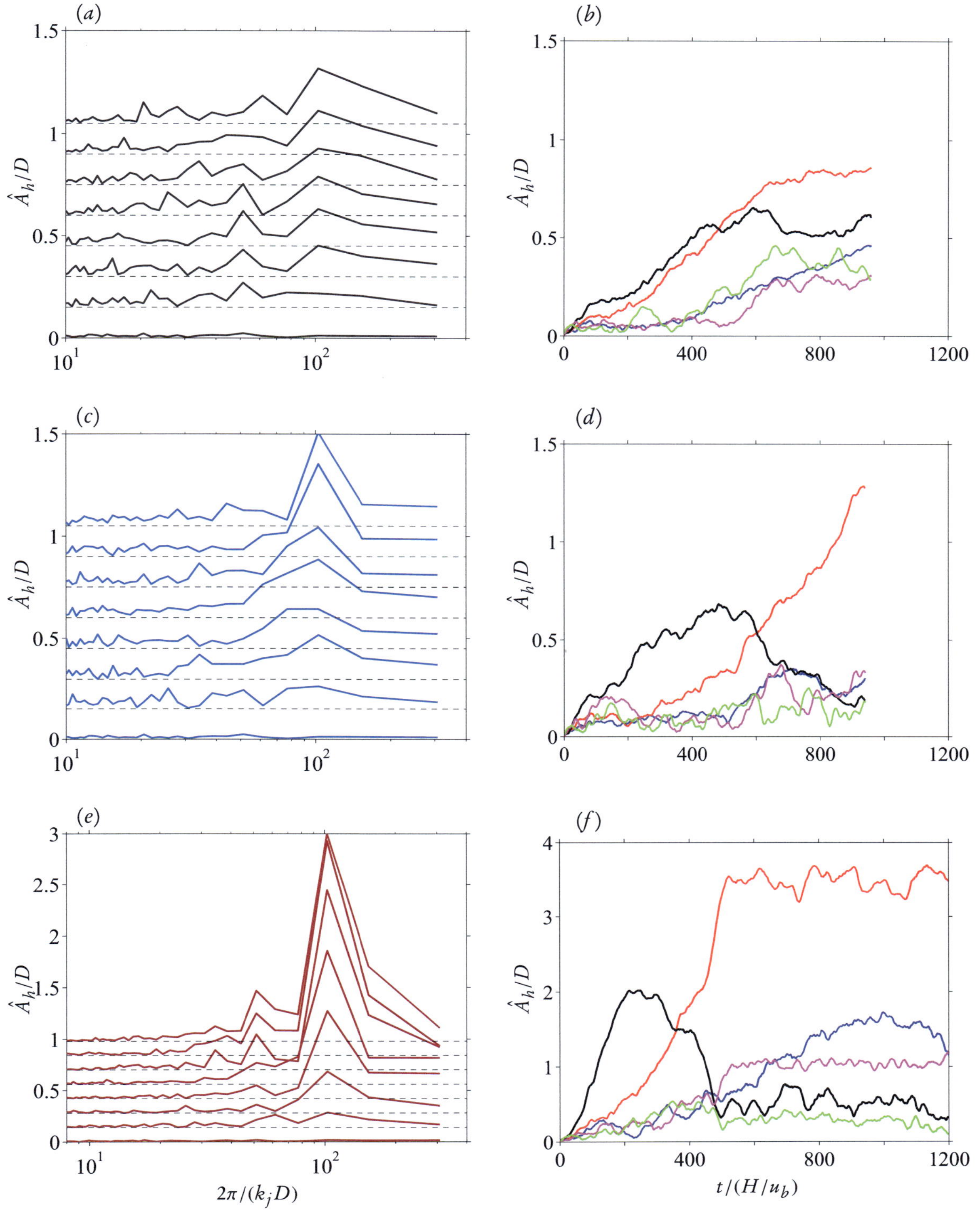

Figure 5.10: (*a,c,e*) Time evolution of the single-sided amplitude spectra for case L2 (—), case L3 (—) and case T1 (—) consecutively. The spectra correspond to the first 250 bulk time units distributed in time equally. Each plot is shifted vertically by a certain amount (indicated by the dashed-lines) for clarity. (*b,d,f*) Time evolution of the amplitude of the most dominant modes for case L2 case L3 and case T1 consecutively. —, $\lambda_1 = L_x$; —, $\lambda_2 = L_x/2$; —, $\lambda_3 = L_x/3$; —, $\lambda_4 = L_x/4$; —, $\lambda_5 = L_x/5$.

growth rate and phase speed of individual modes of $\langle h_b\rangle'_z$, and highlights which wave number, if any, is more dominant out of the entire resolved spectrum. Such an information is the main outcome of theoretical studies of sediment bed instability (Charru et al., 2013).

The space and time varying sediment bed height fluctuation $\langle h_b\rangle'_z$ is transformed into a wavenumber space by applying a discrete Fourier transformation (DFT). Let the periodic streamwise domain length be discretized by N equally spaced grid points located at $x_i = iL_x/N$ for an integer $i \in [0,\ldots,N-1]$. The DFT of $\langle h_b\rangle'_z$ is given by (for an even integer N):

$$\hat{h}_{bj}(t) = \sum_{i=0}^{N-1} \langle h_b\rangle'_z(x_i,t)\exp(-2\pi Iji/N) \quad \forall j = 0,\ldots,N-1 \tag{5.11}$$

where $I = \sqrt{-1}$. The amplitude and phase of the jth harmonic are given as

$$A_j(t) = |\hat{h}_{bj}(t)|/N \tag{5.12a}$$

$$\varphi_j(t) = \operatorname{atan2}(\Im(\hat{h}_{bj}(t)),\Re(\hat{h}_{bj}(t))) \tag{5.12b}$$

respectively. Note that, in the real world, and in our simulations, the resolved wavenumbers are positive-valued, i.e. $k_j = 2\pi j/L_x$ for $j \in [0,\ldots,N/2]$, and thus the amplitude spectrum is symmetric. Moreover, $A_0 = 0$ since $\langle\langle h_b\rangle'_z(x_i,t)\rangle_x = 0$. A single-sided amplitude spectrum is then defined as

$$\hat{A}_h(k_j,t) = 2A_j(t) \quad \forall j = 0,\ldots,N/2. \tag{5.13}$$

The growth rate of the amplitude and the phase speed of individual modes can then be easily evaluated as

$$\sigma_j(t) = \mathrm{d}A_j(t)/\mathrm{d}t \tag{5.14a}$$

$$c_j(t) = -(1/k_j)\mathrm{d}\varphi_j(t)/\mathrm{d}t \tag{5.14b}$$

respectively. It should be note that σ_j is different from, but related to, the growth rate usually defined in most of the theoretical studies. Customarily, the amplitude is defined as $A = A_0\exp(\sigma_j^* t)$ (cf. Charru et al., 2013), and thus $\sigma_j = \sigma_j^*/A_j$.

Figure 5.10 shows the short time evolution of the single sided amplitude spectra corresponding to the pattern-featuring cases for the initial time interval of approximately 250 bulk time units. The figure also shows the evolution of the amplitude of the dominant modes, with wavelengths $\lambda_i = L_x/i$ for $i = 1,2,3,4,5$, covering the entire observation interval. In all the three cases, after approximately 50 bulk time units, it can be observed, that the amplitude of third harmonic ($\lambda_3 = L_x/3$) becomes clearly dominant when compared to the remaining modes. The degree of dominance is more pronounced for cases L3 and T1. For case L2, the second harmonic ($\lambda_2 = L_x/2$) seems to exhibit comparable amplitude to that of λ_3.

The evolution of the amplitude of the dominant modes, which is presented in figure 5.10(*b,d,f*), highlights the subsequent evolution process. These modes contain most of the amplitude of the patterns. For the laminar case L2, the dominant third and second harmonics grow at a comparable rate until $tu_b/H \approx 500$. After this time, the third harmonic is observed to attain a more or less constant amplitude until the end of the simulation. The second harmonic dominates and further grows, but at a milder rate. Moreover, the remaining harmonics, which showed no substantial growth before $tu_b/H \approx 500$, are also observed to gain amplitude

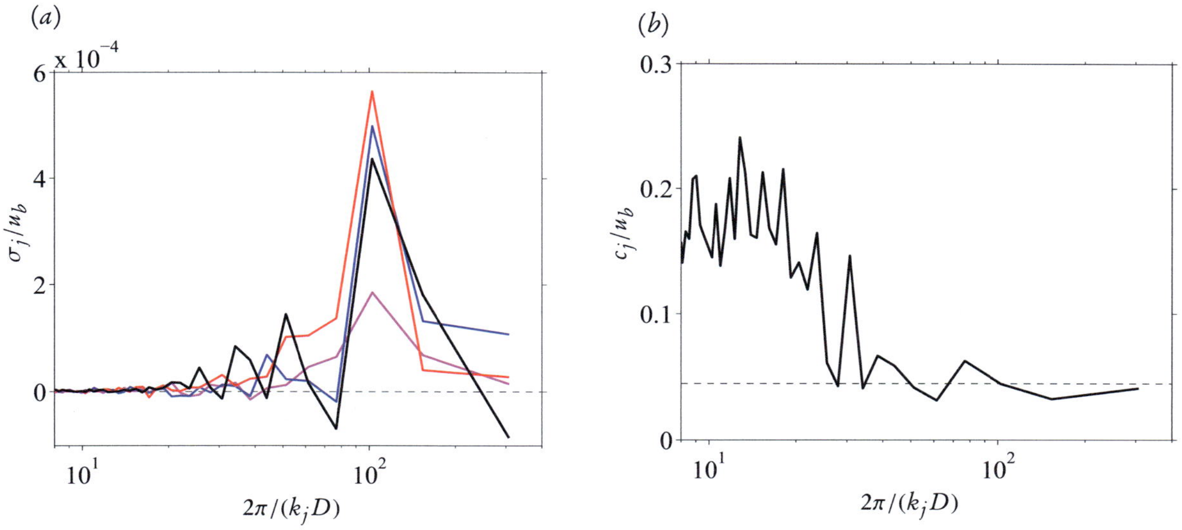

Figure 5.11: Dispersion relation at selected times of case T1. (*a*) Growth rate of amplitude as a function of the wavelength. —, $0 < t/(H/u_b) < 50$; —, $50 < t/(H/u_b) < 100$; —, $100 < t/(H/u_b) < 150$; —, $150 < t/(H/u_b) < 200$; (*b*) Phase speed as a function of wavelength for $150 < t/(H/u_b) < 200$. The horizontal dashed-line corresponds to the mean propagation speed of the patterns in the same time interval.

steadily until the end of the simulation. On the other hand, the amplitude evolution of case L3 shows some differences when compared to that of case L2. Until $tu_b/H \approx 500$, the third harmonic is observed to dominate all the other harmonics. After this time however, its amplitude steadily decreases and loses its dominance to the second harmonic which is observed to grow markedly above the other modes. The remaining harmonics show no clear evolution trend, exhibiting slight increase in amplitude near the end of the simulation.

Turning to the turbulent case T1, the evolution trend resembles that of case L2 somewhat. But at a different length and time scales. The third harmonic rapidly grows in the first approximately 200 bulk time units. Its amplitude is then observed to decrease at almost the same rate as its initial growth, and after $tu_b/H \approx 500$, it loses almost all its amplitude and is observed to not evolve with negligible contribution to the spectra until the end of the simulation. The second harmonic, on the other hand, monotonically increases in the first approximately 500 bulk time units and attains the largest amplitude of approximately $3.5D$. It then levels off and maintains its amplitude until the end of the simulation, clearly dominating the spectra. The first harmonic with a wavelength $\lambda_1 = L_x$ is also observed to gain amplitude steadily from the start of the simulation until $tu_b/H \approx 800$. After this time, its amplitude levels off and even shows a slight decrease near the end of the simulation interval. Overall, the evolution trend of all the cases reveals a non-linear coarsening process of the patterns which has been already evidenced by the auto-correlation function of the bed height.

For completeness, we provide in figure 5.11 a sample dispersion relation at selected initial times during the initial evolution for the turbulent case T1. The figure shows the growth rate of the amplitude and the phase speed of the resolved harmonics. To reduce the level of noisiness, the dispersion relation data has been averaged over intervals of approximately 50 bulk time units. In the initial time interval of 200 bulk time units, the growth rate of the third harmonic is observe to be the highest of these plots, in accordance with the discussion in the previous paragraph. The sample plot of the phase speed, which corresponds to the time interval $150 < t/(H/u_b) < 200$, reveals that the dominant modes propagate with an almost the same phase speed which, in turn, is comparable to the average propagation speed of the patterns in this time interval. The phase speed of the higher wavenumber modes on the other hand, is observed to be much larger than that of the small wavenumber modes, reaching values $c_j \approx 0.2u_b$ for wavelengths in the range $10D < \lambda_i < 20D$, which is approximately four times larger than the mean propagation speed of the patterns.

5.3 Influence of computational box on pattern formation

Ideally, if someone desires to accurately capture the natural selection mechanism of the unstable wavelength λ_c and its subsequent evolution, he or she has to consider a computational domain size which is much superior than the anticipated pattern wavelength. This way, it is insured that a sufficiently high number of modes, with wavelength comparable to λ_c, are contained in the grid. In practice however, due to the very high computational cost of performing DNS of such systems, our simulations are performed by adopting computationally feasible domain sizes. For instance, for the simulations reported so far, $L_x/\lambda_c = O(1)$ and the resolved wavenumbers could be considered marginally discrete when compared to λ_c. In light of this aspect, assessing the influence of the domain size, thus the resolution of the wavenumbers, on the selection of λ_c and its evolution is crucial. Moreover, various theoretical and experimental studies have shown, that there is a lower threshold of the unstable wavelengths (see for instance Franklin and Charru, 2011; Charru, 2006). That means, in a computational domain length which is smaller than a certain cutoff length ($L_x < \lambda_{th}$), an erodible bed will be always stable, which otherwise would have been unstable had L_x been greater than λ_{th}. Determination of λ_{th} is important since the lower bound of unstable modes is believed to be linked to some relevant length scale which controls erodible bed instability (Franklin and Charru, 2011).

In this section we address the influence of L_x on λ_c and determine λ_{th}, by performing six additional simulations of the turbulent case T1, keeping all physical and numerical parameters identical except for the streamwise length of the computational box L_x (cf. table 5.1). These additional simulations are labeled $T1^{2H}$, $T1^{3H}$, $T1^{4H}$, $T1^{6H}$, $T1^{7H}$ and $T1^{48H}$, the superscript indicating the approximate streamwise box length in terms of the fluid height H. For computational cost reasons, these additional simulations do not cover the entire observation interval of case T1. However, the simulated time interval of each simulation is sufficient enough to address the above mentioned aspects. It should be note that case $T1^{48H}$ is a very large-scale simulation with a domain size which is four times larger than that of T1, and with a very large number of spherical particles (approximately 1.1 million in total) representing the mobile bed. This simulation is the first over to break the $O(10^6)$ fully resolved particle barrier.

5.3.1 Lower threshold of amplified pattern wavelength

Our strategy of finding the minimal L_x which will accommodate the lower bound of the unstable wavelengths λ_{th} (for the adopted parameter point) is to perform several numerical experiments in which the streamwise domain length adopted in case T1 ($L_x \approx 12H$) is minimized successively (note that each simulation is performed independently following the simulation start-up procedure detailed in section 5.1.2). The concept is similar to the minimal flow unit of Jiménez and Moin (1991). If a threshold value of L_x is reached, pattern evolution will be hindered and a perturbed bed should in principle be stable even though it is in a regime where instability is expected (for instance at the presently chosen parameter point). Note that determining the lower threshold of L_x for bed instability means determining the lower threshold of the unstable wavelengths. It does not necessarily mean determining the most amplified wavelength out of a range of unstable modes.

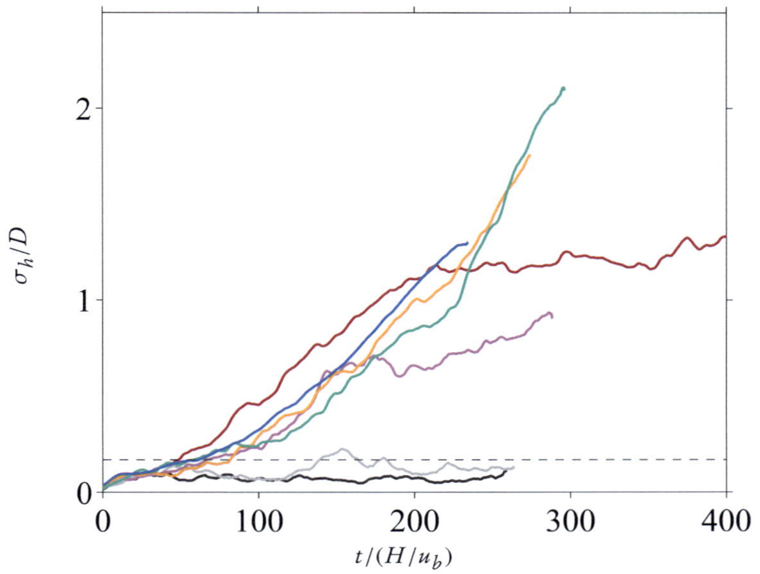

Figure 5.12: Time evolution of the r.m.s. sediment bed height (equation 5.5) normalized by the particle diameter. —, case T1^{2H} —, case T1^{3H} —, case T1^{4H} —, case T1^{6H} —, case T1^{7H} —, case T1 —, case T1^{48H}. The horizontal dashed line indicates the value $\sigma_h/D \approx 0.17$ which corresponds to the random bed fluctuations of featureless flat bed in motion (Coleman and Nikora, 2009).

We have evaluated the r.m.s. sediment bed height fluctuation σ_h (cf. equation 5.5) for each of the cases and used it as a criterion to determine whether a given sediment bed is stable or not. That is, if σ_h remains bounded within the threshold which corresponds to the random bed fluctuations of a flat bed in motion, then the bed is said to be stable. On the other hand, if σ_h exhibits a continuous increases as a function of time, the bed is said to be unstable. Figure 5.12 shows the time evolution of σ_h for all the considered cases. It is seen that, starting form time $tu_b/H \approx 50$, the r.m.s. of the sediment bed height fluctuation grows with time for all the cases with domain length $L_x \gtrsim 102D$ ($L_x \gtrsim 4H$), indicating the chosen streamwise box dimension of these cases is larger than the lower threshold of unstable wavelengths. On the contrary, the evolution of σ_h of cases T1^{2H} and T1^{3H} is observed to show no growth with time except for small fluctuations which are sufficiently bounded by the threshold value which corresponds to a flat bed in motion ($\sigma_h/D \approx 0.17$ as per Coleman and Nikora, 2009). This indicates that, the streamwise box dimension of these cases in which $L_x \lesssim 78D$ ($L_x \lesssim 3H$) is not sufficient to accommodate even the lower threshold of unstable modes. Thus, it could be concluded that the cutoff length scale for pattern formation (for the considered parameter point) is $78D < \lambda_{th} < 102D$, or in terms of the half channel height, $3H < \lambda_{th} < 4H$. Additional simulations with L_x in the range $3H$ to $4H$ would be required in order to determine λ_{th} more accurately.

It should be mentioned that the adopted computational box sizes are chosen such that they are sufficient enough to sustain the turbulence of the flow above the sediment bed. This is confirmed by monitoring the time evolution of the box-averaged turbulent kinetic energy (plots not shown). Note that, in smooth wall channel flows, the minimum spanwise and streamwise box dimensions required in order for turbulence to be self-sustained, are $L_z^+ \approx 100$ and $L_x^+ \approx 350$, respectively (Jiménez and Moin, 1991). The minimum computational box dimensions adopted in our study (corresponding to case T1^{2H}) are substantially larger, i.e. $L_z^+ \approx 735$ and $L_x^+ \approx 490$.

5.3.2 Initially selected pattern wavelength and its subsequent evolution

The time evolution of the sediment bed shown in figure 5.12 also shows that there is a slight difference on the rate of growth among the different 'unstable' cases, although all were carried out at the same imposed parameter values. This is most probably a consequence of the discreteness of the resolved harmonics of a given case and the corresponding selected initial wavelength λ_c. In the reference case T1 and in the small box cases T1^{4H}, T1^{6H} and T1^{7H}, even though the streamwise length of the computational box is larger than the threshold for pattern formation, due to the fact that $L_x/\lambda_{th} = O(1)$, the available wavelengths from which the system has to select, are very discrete, and it is not guarantied that the selected wavelength will be the true critical wavelength which the system would naturally select. On the other hand, case T1^{48H} has a streamwise box length L_x in the range $12\lambda_{th}$ to $16\lambda_{th}$. Thus the initial wavelength selected in case T1^{48H} is expected to be sufficiently close to the true natural wavelength. Thus, in the following, we assess the influence of L_x on the selected wavelength λ_c and its subsequent evolution by comparing each case with the result of case T1^{48H}.

From the time-lapse snapshots of the particle positions of case T1^{48H} shown in figure 5.13, the gradual emergence of spanwise-oriented patterns, which are approximately equally spaced, can be visually witnessed. By simply counting the number of ridges or troughs, it can be easily deduced that the computational domain accommodates eleven ripples with a mean initial wavelength of $\lambda_c = L_x/11 \approx 4.36H \approx 112D$ by the time $tu_b/H \approx 100$. It can be concluded that this value represents the wavelength that would have been selected in a domain with $L_x \gg 48H$ with a confidence interval sufficiently smaller than $\pm 10D$ since the neighboring resolved harmonics around λ_c have a wavelength $\lambda = L_x/12 \approx 102D$ and $\lambda = L_x/10 \approx 123D$. At first impression, λ_c selected in case T1^{48H} is slightly larger than that selected in the reference case T1 ($\lambda_c \approx 102D$) highlighting the effect of L_x on λ_c. A similar qualitative description of the remaining cases is performed and a sample particle position visualization at a particular instant ($tu_b/H \approx 200$) is shown in figure 5.14. Once again, several conclusion can be made from these images. In accordance with the results in the previous section, no patterns are observed for cases T1^{2H} and T1^{3H}. However, one can observe, that it exhibits distinct streamwise aligned alternating ridges and troughs, although the sediment bed is essentially flat when averaged in the spanwise direction. Such an organization of particles believed to be linked to the streamwise-aligned near-wall turbulent structures which are responsible for the non-homogeneous spatial distribution of particles. As a result of the very small box length, no spatial de-correlation of the turbulent structures (in the streamwise direction) is expected which results in the observed highly regular ridge-trough pattern. On the other hand, cases T1^{4H}, T1^{6H} and T1^{7H} allow the formation of spanwise-oriented patterns. These cases exhibit a common feature; the dominant wavelength selected in all three cases is equal to the streamwise box length L_x. This means that the selected wavelength is equal to $\lambda_c \approx 102D$ in case T1^{4H}, $\lambda_c \approx 154D$ in case T1^{6H} and $\lambda_c \approx 179D$ in case T1^{7H}, highlighting the substantial influence of L_x on λ_c in these cases when compared to cases T1 and T1^{48H}.

A more quantitative description of the selection of the initial wavelength and its subsequent evolution can be provided by computing the mean wavelength of the pattern using equation (5.7). The result is presented in figure 5.15, which compares the time evolution of the mean wavelength of T1^{48H} with all the other bedform-featuring cases. In figure 5.15(a), T1 and T1^{48H} are compared. The initial bed instabilities in both cases are seen to be almost the same. Starting from time $tu_b/H \approx 100$ however, λ_h of case T1^{48H} is observed to attain a value corresponding to the eleventh harmonic with wavelength $L_x/11 \approx 112D$ and is observed to increase further exhibiting a relatively constant growth rate until the end of the simulated interval, while λ_h of case T1 remains constrained to the third harmonic ($L_x/3 \approx 102D$) during the same time interval. As has

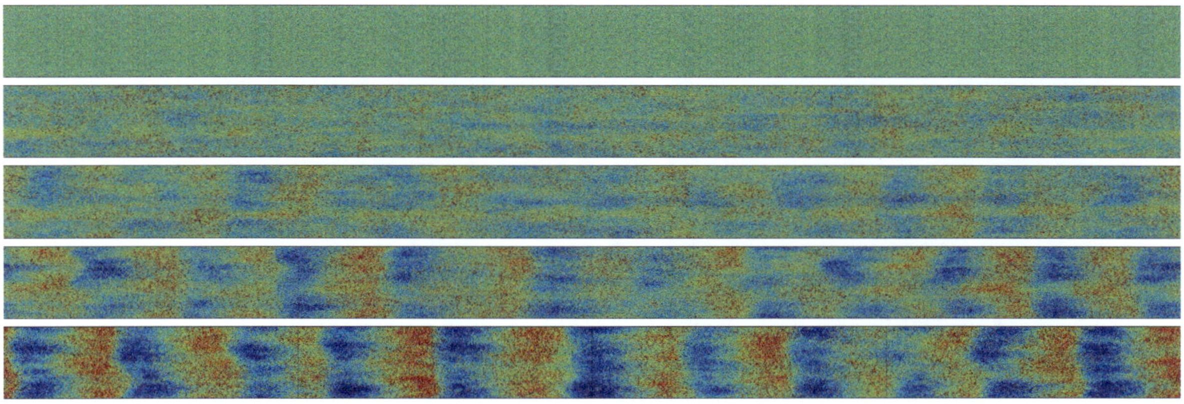

Figure 5.13: Top view of the three-dimensional instantaneous particle positions of case $T1^{48H}$. The wall-normal dimension is visualized by coloring each particle according to its wall-normal location (increasing from blue to red). From top to bottom, snapshots correspond to time $tu_b/H = \{0,50,100,150,200\}$.

been presented in figure 5.7, the mean wavelength in case T1 then jumps to the harmonic with wavelength $L_x/2 \approx 154D$ at relatively increased growth rate in the time interval $250 \gtrsim tu_b/H \gtrsim 500$. This was attributed to the unavailability of resolved harmonics in between. Case $T1^{48H}$, on the other hand, possesses additional three resolved harmonics which allows λ_h to grow relatively steadily. Thus, it can be concluded that the domain length adopted in case T1, which accommodated three initial ripple wavelengths, is marginally sufficient in determining the critical wavelength, but it is too small to capture the subsequent steady evolution of the selected wavelength. Turning to cases $T1^{7H}$, $T1^{6H}$ and $T1^{4H}$, it is seen in figure 5.15(*b*-*d*) that the chosen domain length in these cases strongly constrains the selected initial wavelength. In all three cases, the mean wavelength jumps withing the first 100 to 150 bulk time units to the maximum possible wavelength, i.e. $\lambda_h = L_x$ and evolves constrained to this value.

5.4 Summary and conclusion

We have performed direct numerical simulation of the flow over an erodible bed of spherical sediment particles both in the laminar and turbulent regimes. In total, ten separate simulations were performed covering four sets of parameter points. The adopted values of the physical and numerical parameters are presented in table 5.1. One of the laminar cases evolved without the formation of patterns while the other two laminar cases lead to the formation of 'small dunes'. The simulation performed under turbulent flow conditions exhibited 'vortex dunes', consistent with the regime classification of Ouriemi et al. (2009a).

The statistically two-dimensional fluid-bed interface, i.e. the sediment bed height, has been extracted from the spanwise-averaged solid volume fraction prescribing a threshold value. The spatial and temporal variation of the sediment was then analyzed in order to investigate the time evolution of characteristic streamwise and wall-normal dimensions of the emerged patterns as well as their propagation velocity. In the laminar cases, the instabilities of the sediment bed start to emerge within the first 50 bulk time units, after which the amplitude of the formed 'small dunes' exhibited a gradual but monotonous growth until the end of the simulated observation interval, showing no sign of reaching a steady state. On the other hand, the amplitude of the 'vortex dunes' initially grew at a rate one order of magnitude larger than that of the 'small dunes' (within the first approximately 200 bulk time units). Subsequently, constrained by the discreteness of the resolved

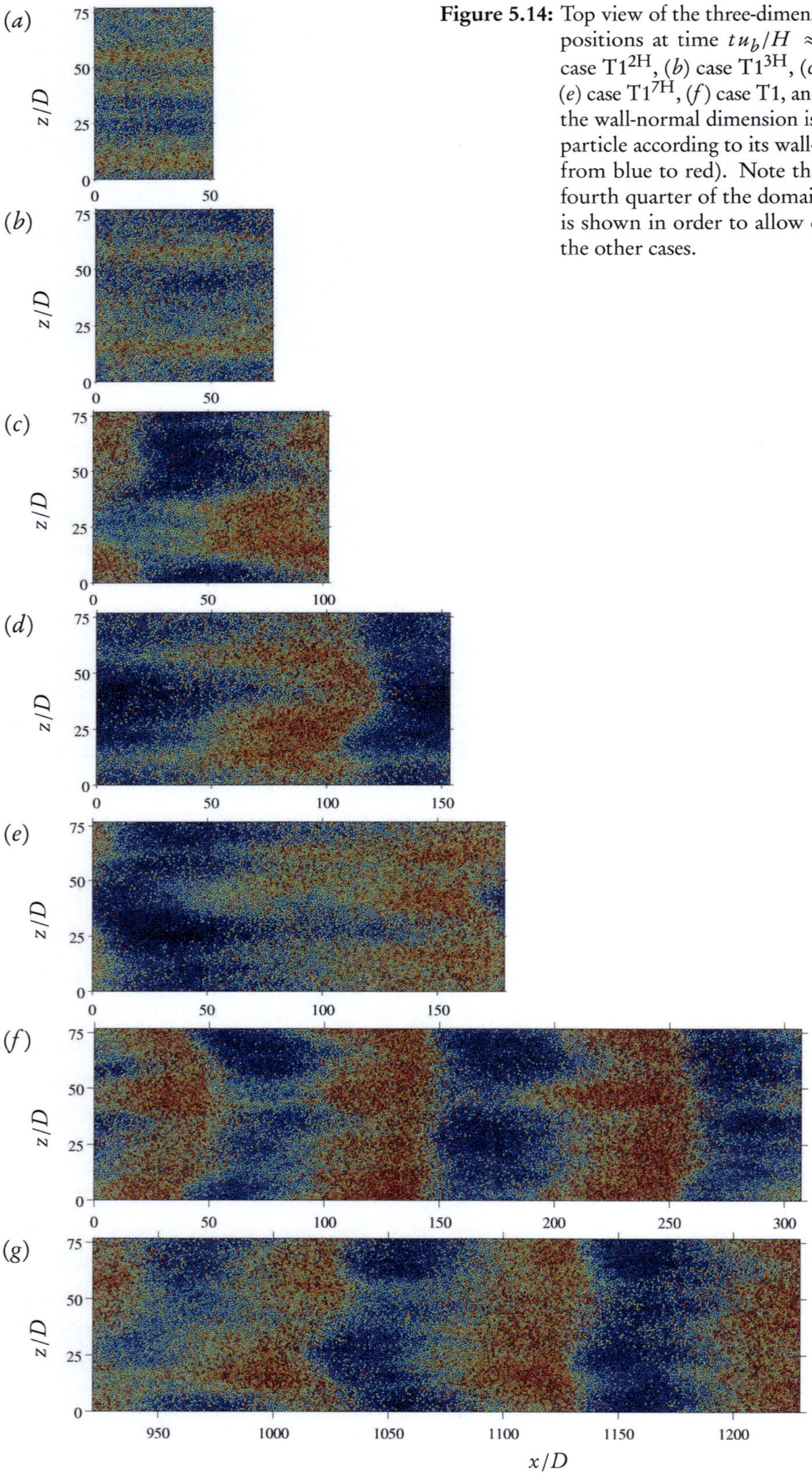

Figure 5.14: Top view of the three-dimensional instantaneous particle positions at time $tu_b/H \approx 210$ corresponding to (*a*) case T1^{2H}, (*b*) case T1^{3H}, (*c*) case T1^{4H}, (*d*) case T1^{6H}, (*e*) case T1^{7H}, (*f*) case T1, and (*g*) case T1^{48H}. In all plots, the wall-normal dimension is visualized by coloring each particle according to its wall-normal location (increasing from blue to red). Note that for case T1^{48H}, only the fourth quarter of the domain ($921.6D < x < 1228.8D$) is shown in order to allow direct visual comparison to the other cases.

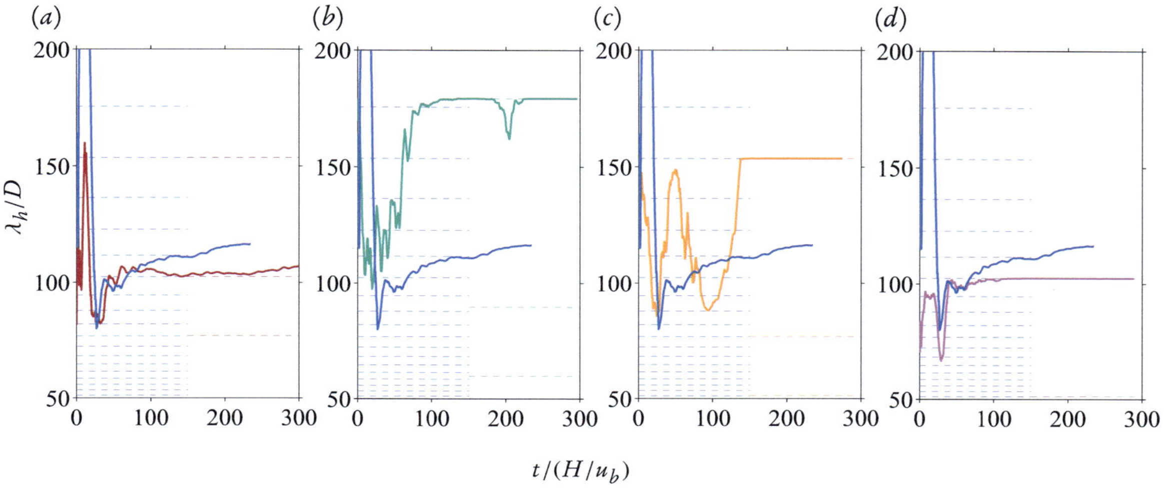

Figure 5.15: Time evolution of the mean wavelength of the bed height normalized with the particle diameter. The dashed lines indicate the wavelengths of the first few harmonics resolved in the domain of the respective case. The plots show the evolution of λ_h of case $T1^{48H}$ compared to that of: (*a*), case T1; (*b*), case $T1^{7H}$; (*c*), case $T1^{6H}$; (*d*), case $T1^{4H}$. Color coding similar as in figure 5.12

harmonics contained in the computational box, it exhibited an intermittent behavior of alternating small and large growth rate intervals throughout the simulated duration, and finally, in the last approximately 400 bulk time units, it evolved almost steadily and showed no growth. Overall, the growth rate of the amplitude of the 'small dunes' and 'vortex dunes' is found out to be in good agreement with the available experimental data (see e.g. Langlois and Valance, 2007; Ouriemi et al., 2009a).

In both the laminar and turbulent cases, the wavelength of the patterns, which was statistically determined from the two-point correlation function of the sediment bed height fluctuations, attained values in the range $\lambda_h \approx 100D - 150D$ ($\lambda_h \approx 4H - 6H$), in the considered observation interval. Again, these values are consistent with finding of the previously cited experimental works. Similar to the the growth of the amplitude, a non-monotonous growth of the wavelength was also observed. Such an evolution trend is believed to be a consequence of the relatively small streamwise length of computational box adopted. That is, the system is constrained to the available discrete harmonics and is forced to exhibit the observed intermediate plateau regions. The non-linear pattern coarsening process was further studied by investigating the evolution of the pattern morphology, namely the aspect ratio and the degree of asymmetry of the bedforms. Generally, the aspect ratio increases from an initial negligible value to a final value of approximately 1.5% for the 'small dunes' and approximately 3.7% for the 'vortex dunes'. Moreover, both 'small dunes' and 'vortex dunes' evolved from an initial symmetric shape towards the well known asymmetrical shape of dunes.

The propagation speed of the patterns was determined from the shift of the maximum of the two-point/two-time correlation of the fluid-bed interface fluctuation. It turns out that the patterns exhibit a generally decreasing drift velocity as a function of time. This shows, that the propagation speed of the pattern is inversely proportional to the dune dimensions. This finding is consistent with previous studies (cf. Führböter, 1983; Betat et al., 2002; Coleman et al., 2003; Ouriemi et al., 2009a).

The turbulent case simulation is repeated several times by keeping all relevant parameter values identical but minimizing the computational domain in the streamwise direction in order to find the minimal box length which will accommodate the lower bound of the unstable pattern wavelength at the instants of the

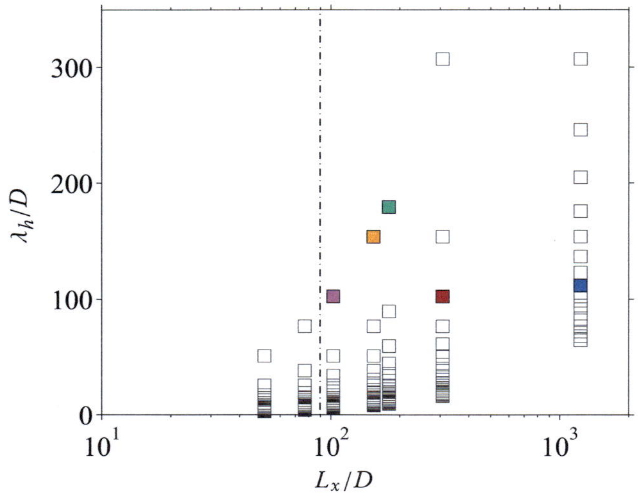

Figure 5.16: Initially selected pattern wavelength as a function of the streamwise length of the computational box for cases ■, case $T1^{4H}$; ■, case $T1^{6H}$; ■, case $T1^{7H}$; ■, case T1; ■, case $T1^{48H}$. The open-square symbols (□) indicate the wavelengths of the first few resolved harmonics in all seven simulations. Note that in cases $T1^{2H}$ and $T1^{3H}$, no formation of patterns was observed.

initial bed instability. The concept is similar to the minimal flow unit of Jiménez and Moin (1991). If the streamwise box length is below a threshold value, pattern evolution will be hindered and a perturbed bed will remain stable. For the chosen parameter point, it was found out, that a computational box with a streamwise dimension $L_x \lesssim 78D$ was not sufficient to accommodate any of the unstable modes hindering the growth of sediment patterns. On the contrary, a box with $L_x \gtrsim 102D$ accommodated at least one unstable mode. This observation indicates that the cutoff length for pattern formation lies in the range $78D < \lambda_{th} < 102D$, or in terms of the half channel height, $3H < \lambda_{th} < 4H$.

The influence of the computational domain size on the selection and evolution of the most amplified wavelength was further studied by performing one simulation adopting a domain size with a streamwise length which is much larger than the anticipated initial pattern wavelength. The box was able to accommodate eleven initial wavelengths and allowed the determination of the most amplified initial wavelength to be determined with sufficient accuracy. It turns out that the most amplified wavelength corresponds to the resolved harmonic with wavelength $\lambda_c \approx 112D \approx 4.36H$. Figure 5.16 summarizes the influence of the computation domain length on the initially selected pattern wavelength. It can be concluded that a computational domain length with $L_x/\lambda_c = O(1)$, due to the discreteness of the resolved harmonics, severely constrains the natural selection mechanism of the initial pattern wavelength.

Chapter 6

Fluid flow and particle motion over subaqueous patterns in channel flow

In this chapter, we extend the investigation made in the previous chapter, focusing on the characterization of the temporally and spatially varying flow field over the subaqueous bedforms, the main actor in their formation. Equal focus is also given to the characterization of the associated individual grain motion, the main process by which an unstable sheared sediment bed give way to the formation and evolution of the bedforms. In particular, we endeavor to contribute to the ongoing quest of answering the following important questions:

- How can the space- and time-dependent turbulent flow over bedforms be characterized?
- What does the spatial structure of the turbulent flow and particle motion look like?
- How does the basal shear stress correlate with the morphology of the sediment bed?
- How do sediment grains respond, individually or collectively, to the flow field?
- How does the evolving sediment bed correlate with the dynamics of sediment grains?

For this purpose, the time evolution of bulk statistics of the flow, such as the turbulent kinetic energy and total bottom shear stress is discussed in detail. The time evolution of the particle volumetric flow rate and its relation with the time-varying bottom shear stress is also discussed. Furthermore, for the purpose of accurately correlating the turbulent flow field and particle motion with the bedforms, an averaging procedure is presented which takes into account the spatial and temporal variability of the sediment bed. The technique is then exploited to analyze the DNS data corresponding to the turbulent case T1 from the previous chapter.

The physical and numerical parameters of the cases has been provided in table 5.1 of the previous chapter.

6.1 Time evolution of bulk statistics

We have seen in the previous chapter that the evolution of the sediment is in a transient state, except for the turbulent case T1 in which the bed exhibited an approximately steady-state interval near the end of the simulation. The steady-state behaviour could be either as a result of the system reaching a developed state due

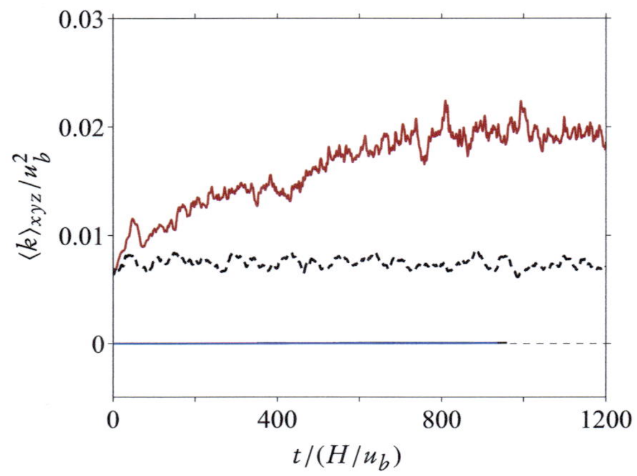

Figure 6.1: Time evolution of box-averaged instantaneous turbulent kinetic energy —,L2; —,L3; —,T1; - - -, single-phase flow at a bulk Reynolds number $Re_b = 2870$ which is computed (by a pseudo-spectral method) in a computational domain size which is identical to the domain occupied by the fluid region (of height h_f) in T1.

to a physical finite-size effect, for instance the influence of the upper boundary (Charru et al., 2013) or as a result of the limitation of the size of the computational box (discreteness of harmonics preventing the system to jump to the next mode). It is expected that the spatially and temporally varying sediment bed is mutually interacting with the driving flow, altering the properties of the latter continuously. The degree of influence of the sediment bed on the driving flow could however be different between the laminar 'small dune' cases and the turbulent 'vortex dune' cases. In the former, the maximum perturbations of the sediment bed attained are observed to be much smaller when compared to the fluid height. Thus the mean flow statistics are expected to be not substantially affected by the evolution of the bed in these cases. In the turbulent case on the other hand, the amplitude of the sediment patterns reached approximately 6 particle diameters or reaching up to 24% of the half channel height. (cf. figure 5.7 of chapter 5) and this is expected to influence the flow field considerably. Thus the mean flow statistics for the turbulent case is expected to evolve from a statistically one-dimensional state, during the initiation of the simulations, towards a time varying two dimensional flow field, in the vicinity of the sediment bed in particular. In this section, we investigate the time-variation of the bulk flow characteristics, such as the mean velocity and mean shear stress.

6.1.1 Box-averaged energy of fluid velocity fluctuations

The time evolution of the total energy contained in the fluid velocity fluctuations with respect to the instantaneous mean velocity profile, averaged over wall-parallel planes, is given by

$$\langle k\rangle_{xyz}(t) = \frac{1}{2}\left[\langle u_{f,\alpha}u_{f,\alpha}\rangle_{xyz}(t) - \langle\langle u_{f,\alpha}\rangle_{xz}\langle u_{f,\alpha}\rangle_{xz}\rangle_{y}(t)\right], \tag{6.1}$$

where summation is implied over the repeated index α. This quantity is depicted in figure 6.1 for the pattern-featuring cases L2, L3 and T1. Additionally, the figure shows the evolution of the turbulent kinetic energy corresponding to a smooth-channel single phase flow ($Re_b = 2870$; $Re_\tau = 183$) for the purpose of comparison with case T1. The bulk statistics of this reference data has been previously presented by Kidanemariam et al. (2013). Note that $\langle k\rangle_{xyz}$ represents not only the box-averaged turbulent kinetic energy but also contributions coming from the spatial variability of the mean flow as a result of the evolving features of the sediment bed.

For the laminar cases L2 and L3, $\langle k\rangle_{xyz}$ is practically zero, indicating that the continuously evolving patterns have almost no influence on the characteristics of the mean flow. The flow stays essentially one-dimensional which only depends on the wall-normal direction, at least in the simulated observation interval. The turbulent case T1, on the other hand, exhibit a substantial increase of $\langle k\rangle_{xyz}$ from its initial value. At the start of the observation interval (the time particle are first released to freely move), $\langle k\rangle_{xyz}$ has comparable value to that of the reference single phase flow, i.e. it is hydraulically smooth with $D^+ \approx 10$. After this time, fluid fluctuations immediately start to increase exhibiting a continuous growth of the energy until $tu_b/H \approx 800$ reaching values more than twice the initial value. Then, $\langle k\rangle_{xyz}$ is observed to show a more or less constant value until the end of the simulation. It should be noted however that the evolution of $\langle k\rangle_{xyz}$ of case T1, as is presented in figure 6.1, does not explicitly represent the energy of the actual turbulence. This is because of the increasing deficiency of the one-dimensional mean flow assumption near the sediment bed as the roughness elements, i.e. the bedforms, evolve in time and the mean flow becomes increasingly two-dimensional (Yalin, 1977). This will be addressed in more detail in later sections.

6.1.2 Imposed pressure gradient and bottom shear stress

As was mentioned in the previous chapter, the pressure gradient which drives the flow is adjusted at each time step in order to maintain a constant flow rate. The instantaneous value of the pressure gradient can be determined from the balance between the driving volume force (the imposed pressure gradient itself) and resisting forces, taking into account the incompressibility constraint. As a result of the free-slip boundary condition at the top of the domain, the resistance is entirely due to the total hydrodynamic and inter-particle collision forces exerted by the stationary particles located beneath the mobile sediment layer, and thus equation 4.18 in chapter 4 reduces to

$$\frac{\mathrm{d}p}{\mathrm{d}x}(t) = -\frac{1}{L_x L_y L_z}\sum_{l=1}^{N_p}\left(F_{p,x}^{H(l)}(t) + F_{p,x}^{C(l)}(t)\right)I_{stat}^{(l)} . \tag{6.2}$$

Figure 6.2(a) shows the evolution of the pressure gradient computed using relation (6.2) for the turbulent case T1, for the entire observation interval. We will discuss the normalization below, but at first glance the figure reveals, that an increasing driving force is required to maintain the desired constant flow rate which indicates that the shear stress exerted by the flow on the evolving sediment bed is increasing with time. The evolution does not show a monotonically increasing trend but exhibits alternating high growth rate regions and plateau regions. This trend closely correlates with the evolution of the sediment bed, for instance the r.m.s. sediment bed fluctuations shown figure 5.7 of chapter 5.

Recalling that, in a fully-developed smooth-wall single phase channel flow, the mean pressure gradient is the slope of the total shear stress profile, $\langle \mathrm{d}p/\mathrm{d}x\rangle$ is related to the bottom wall friction as (cf. section 2.1.2):

$$\left\langle\frac{\mathrm{d}p}{\mathrm{d}x}\right\rangle = -\frac{\rho_f u_\tau^2}{H} \tag{6.3}$$

where the total shear stress is the sum of viscous shear stress, Reynolds shear stress and stress contribution from the fluid-particle interaction (Uhlmann, 2008; Kidanemariam et al., 2013). In the configurations considered in this study however, due to the roughness introduced as a result of the evolving sediment bed, a 'horizontal bottom wall' does not exist. Nevertheless, a virtual wall could be defined to be located at a wall-normal

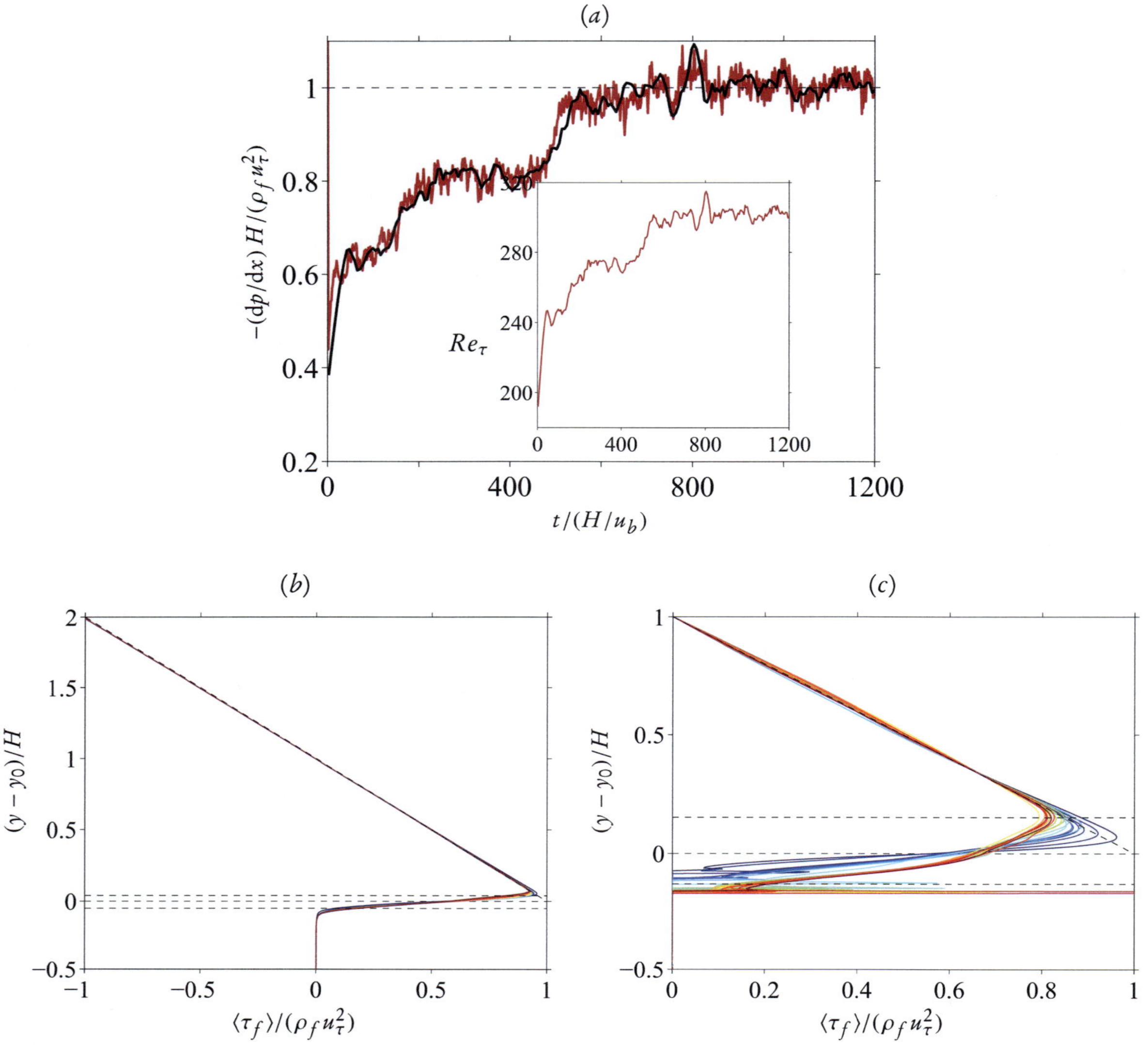

Figure 6.2: (a) Time evolution of the imposed pressure gradient which maintains a constant flow rate q_f for case T1 (—) and the quantity $\rho_f u_\tau^2/H$ corresponding to the friction velocity and half-channel height computed in small time intervals of 25 bulk time units (—). Both quantities are normalized by the asymptotically steady values of u_τ and H. The inset shows the evolution of the corresponding friction velocity based Reynolds number. The evolution of selected wall-normal profiles of the total fluid shear stress for case L3 and T1 (computed in the above small time intervals) are shown in figures (c) and (d) respectively. The profiles are normalized by the value of the bottom shear stress corresponding to their respective time interval. The horizontal dashed lines represent the location of the fluid-bed interface as well as the extents of global extrema of the sediment bed height. The profiles are colored according to time; dark blue corresponds to time interval near the beginning of the simulation and dark red to time interval near the end.

location $y_0 = \langle h_b \rangle_{zx}(t)$. A time varying quasi-instantaneous mean bottom friction could then be defined by extrapolating the total fluid shear stress

$$\langle \tau_f \rangle(y) = \rho_f \nu \frac{\mathrm{d}\langle u_f \rangle}{\mathrm{d}y}(y) - \rho_f \langle u'_f v'_f \rangle(y) \tag{6.4}$$

in the clear fluid region sufficiently above the sediment bed roughness (where the streamlines of the mean flow are sufficiently uni-directional and $\langle\tau_f\rangle$ varies linearly as a function of y) down to the location of the virtual wall. For this purpose, the entire observation time interval is decomposed into smaller intervals of approximately 25 bulk time units each. Flow statistics are then computed in these time intervals by performing standard averaging of a channel flow. The duration of the time intervals is chosen such that it is much smaller than the evolution time scales of the patterns. Sample profiles of $\langle\tau_f\rangle$ for case L3 and T1 for different times spanning from the start of the observation interval to the end are shown in figures 6.2(*b*,*c*). In the clear fluid region sufficiently above the sediment bed, the profiles exhibit an approximately linear variation as a function of wall-normal distance. In regions near the bed however, $\langle\tau_f\rangle$ of case T1 appreciably deviates from the linear variation and the degree of deviation increases with time. $\langle\tau_f\rangle$ of case L3, on the other hand, evolves showing only marginal changes. Note that the second term in the RHS of equation (6.4) is zero for the laminar cases. Furthermore, the evolution of the pressure gradient for the laminar cases is observed to be not affected by the evolution of the sediment bed, attaining a constant value thorough out the observation interval (plots omitted).

The evolution of the mean pressure gradient for case T1, computed by relation (6.3), where the value of the friction velocity is approximated as discussed above, is also shown in figure 6.2(*a*). As expected, a good match between the actual and the estimated values of the pressure gradient is observed. Furthermore, as was shown in figure 5.4 of chapter 5, although the sediment bed fluctuations increase as a function of time, the average sediment bed height $\langle h_b\rangle_{zx}$ is relatively constant for the entire observation interval except for small duration during the start of the simulation. Thus the increase in the shear stress is entirely coming from the increase of the amplitude of the evolving sediment patterns. Incidentally, as is shown in the inset of figure 6.2(*a*), the friction velocity based Reynolds number correspondingly increases from an initial value of $Re_\tau \approx 190$ to a final value of $Re_\tau \approx 300$. Note that the initial value of Re_τ closely matches that of a rough-wall channel flow at comparable bulk Reynolds number and roughness height Chan-Braun et al. (2011).

The evolution of pressure gradient $\mathrm{d}p/\mathrm{d}x$ of the turbulent case T1 (and that of $\langle k\rangle_{xyz}$ as well) exhibit a steady-state region in the time interval greater than $tu_b/H \approx 800$. A mean friction velocity determined in this interval is used for normalization in figure 6.2(*a*) as well as in determination of relevant parameters presented in table 5.1 of chapter 5.

6.1.3 Fluid velocity

Figures 6.3(*a*,*b*) show the time evolution of the wall-normal profile of the mean streamwise fluid velocity $\langle u_f\rangle$, each computed over the small time intervals mention previously, for both the laminar and turbulent cases. The profiles are normalized with the bulk velocity and the half channel height. The fluid velocity profiles of the laminar cases are observed to vary with time only slightly, confirming that the evolution of the sediment bed has little influence in these cases. On the contrary, $\langle u_f\rangle$ of the turbulent case T1, noticeably evolve from an initial profile which is close to that of the single phase smooth channel flow. It exhibits substantial non-zero velocity below the location of the virtual wall, a velocity deficit in the region above y_0 up to $y \approx y_0 + 0.5H$ and a surplus velocity in the outer region above $y \approx y_0 + 0.5H$, with an overall 'flattening' effect of the profile. The degree of flattening is observed to increase with time.

Figure 6.3(*c*) shows the velocity profiles of the turbulent case scaled in wall-units, i.e. the quasi-instantaneous friction velocity and the viscous length scale ν/u_τ are used to normalize the velocity profiles. The corresponding profile for the single phase smooth channel flow is also shown in the figure. It is clearly seen that in

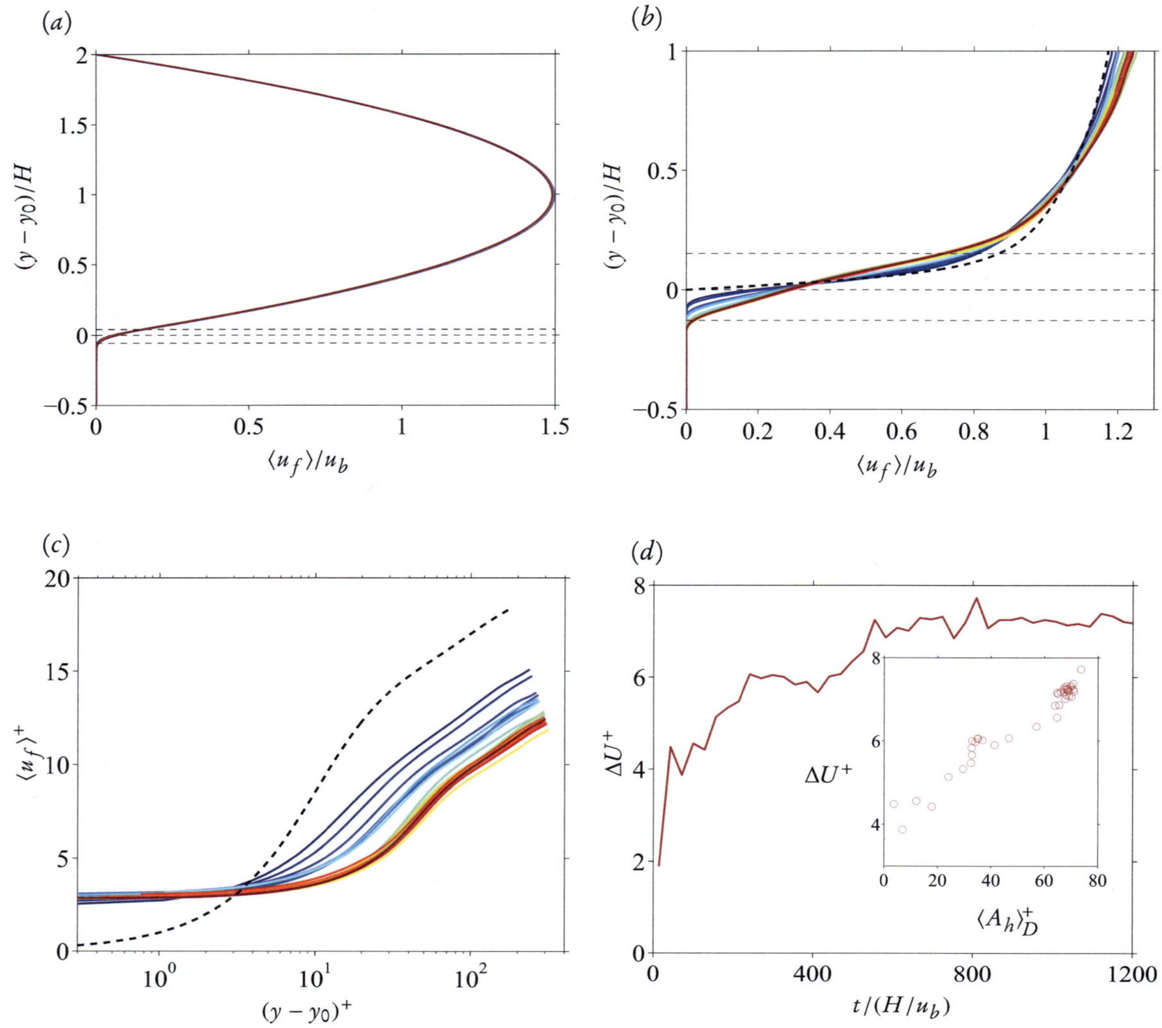

Figure 6.3: Time evolution of the wall-normal profile of the mean streamwise fluid velocity for (*a*) case L3 and (*b*) case T1. The horizontal dashed lines represent the location of the fluid-bed interface as well as the extents of global extrema of the sediment bed height. The profiles are colored according to time; dark blue corresponds to time interval near the beginning of the simulation and dark red to time interval near the end. (*c*) Velocity profiles, in wall units, for case T1. The profiles are normalized by the value of the friction velocity and viscous length scale corresponding to their respective time interval. (*d*) Time evolution of the roughness function ΔU^+. The inset shows the relationship between the roughness function and the roughness height of the patterns.

the region sufficiently above the sediment bed, the velocity profile shifts downwards as a function of time. The alteration to the mean velocity as the sediment bed evolves is mainly as a consequence of the increase of the bottom friction as a result of increasing roughness height of the patterns. In the log region, it is well known, that the velocity profile of channel flow over a rough bottom exhibits a vertical shift when compared to the profile over a smooth wall. The amount of this shift, ΔU^+, which is termed the roughness function, is directly proportional to the roughness height and is usually used to characterize roughness (Jiménez, 2004; Pope, 2000; Chan-Braun et al., 2011; Yuan and Piomelli, 2014). The time evolution of ΔU^+, computed with reference to the smooth channel profile, is presented in figure 6.3(*d*). The evolution trend is consistent with the evolution of the bottom friction. Of particular interest to engineering applications would be to quantify the effect of roughness as a function of the wall-normal dimension of the evolving patterns. The inset in figure 6.3(d) shows, for instance, ΔU^+ as a function of the Reynolds number defined by the mean amplitude of the patterns and the quasi-instantaneous friction velocity $\langle A_h \rangle_D^+(t) = \langle A_h \rangle_D(t) u_\tau(t)/\nu$. The figure seems to suggest an approximately linear relationship between the two quantities, at least in the considered observation interval. It should be noted, however, that the definition of ΔU^+ is subject to the definition of the location of the virtual wall. Different magnitudes of ΔU^+ would result, had the virtual wall been defined to be located at a different location. Nevertheless, the conclusion regarding the linear variation of ΔU^+ as a function of the roughness elements, is expected to hold. Rigorous analysis of this aspect is however not the central focus of the thesis and is left for future investigation.

6.1.4 Volumetric particle flow rate

The time evolution of the volumetric particle flow rate is computed as follows. First the computational domain is decomposed into discrete streamwise-orthogonal bins having streamwise width of Δ_x and spanning the entire wall-normal and spanwise extents of the domain. Let x_i denote the streamwise location of the center point of bin i. A spatially varying instantaneous volumetric particle flow rate (per unit span), $\langle q_p \rangle_z$, is then defined as:

$$\langle q_p \rangle_z(x_i,t) = \frac{\pi D^3}{6 L_z \Delta_x} \sum_{l=1}^{N_p} u_p^{(l)}(t) I_i^{(l)}(t) , \tag{6.5}$$

where $u_p^{(l)}(t)$ is the streamwise component of the instantaneous velocity of particle l at time t and $I_i^{(l)}$ is an indicator function which tells whether the center position of particle l resides in the domain occupied by bin i viz.

$$I_i^{(l)}(t) = \begin{cases} 1 & \text{if} \quad (i-1)\Delta_x \leq x_p^{(l)}(t) < i\Delta_x , \\ 0 & \text{else} . \end{cases} \tag{6.6}$$

A bin width of $\Delta_x = 5D$ is adopted in our analysis. An instantaneous volumetric particle flow rate is then simply a streamwise average of $\langle q_p \rangle_z$viz.

$$\langle q_p \rangle_{zx}(t) = \langle \langle q_p \rangle_z(x_i,t) \rangle_x , \tag{6.7}$$

which is identical to equation (4.8) in chapter 4.

Figure 6.4(a) shows the time evolution of the volumetric particle flowrate for all the laminar cases L1, L2 and L3 normalized by the viscous scale $q_v = Ga^2\nu$ (shown by the solid thick lines). The corresponding quantity for the turbulent case T1, normalized by the inertial scale

$$q_i = \sqrt{(\rho_p/\rho_f - 1)|\mathbf{g}|D^3} = U_g D , \tag{6.8}$$

is shown in figure 6.4(b). The normalization scales come from dimensional analysis and are commonly used in the sediment transport community (see e.g. Ouriemi et al., 2009b). The graphs also show the evolution of the instantaneous maximum and minimum values (over the entire streamwise extent) of the particle flow rate, shown as dashed lines. As is expected, the instantaneous mean, as well as the extreme values, of the featureless case L1 (flat bed in motion regime) do not exhibit a change with time except for a small initial transient interval. The same trend is observed in cases L2 and L3 (small-dune regime) regarding $\langle q_p \rangle_{zx}$. In these cases, the continuous growth of the pattern amplitudes seems to have only a mild influence upon the evolution of

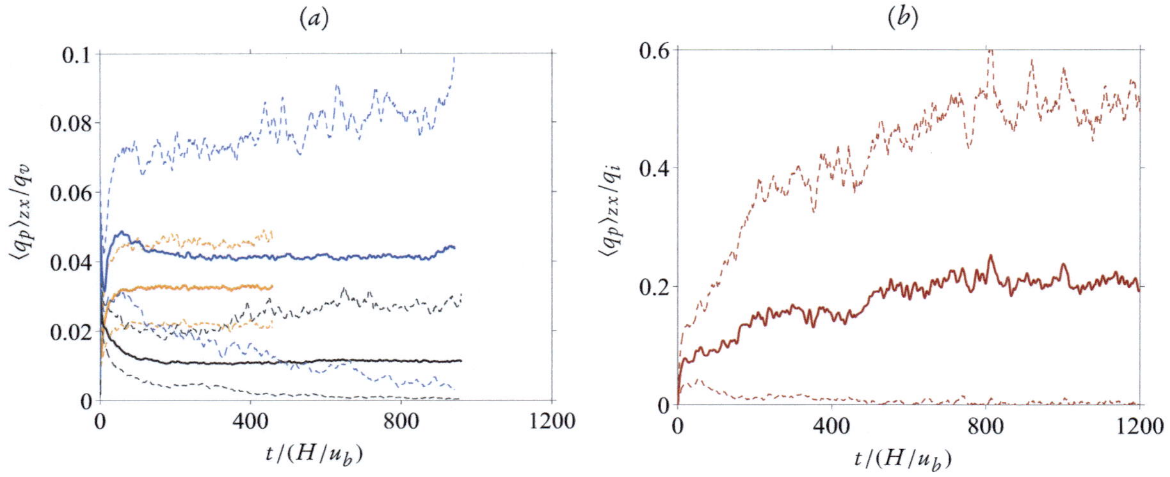

Figure 6.4: (*a*) Time evolution of the instantaneous volumetric particle flowrate $\langle q_p \rangle_{zx}$ for the laminar cases L1(—), L2 (—) and L3(—). The viscous scale $q_v = Ga^2 \nu$ is used for the purpose of normalization. Additionally, the two dashed lines in each case indicate the respective intantaneous global minimum and maximum values of $\langle q_p \rangle_z$. (*b*) The same quantity of the turbulent case T1 but normalized by the inertial scale $q_i = U_g D$.

the spatially averaged particle flow rate $\langle q_p \rangle_{zx}$ as it is observed to attain an approximately constant value for most of the observation interval. On the other hand, it can be observed, that the maximum value continue to grow while the minimum value asymptotically approaches to zero, until the end of the simulations. This trend is consistent with the increase in the amplitude of the propagating patterns. That is, as the small dunes evolve, the spatial variability of the particle flow rate increases. The turbulent vortex-dune regime shows a somewhat different evolution process, namely, $\langle q_p \rangle_{zx}$ exhibits a small but noticeable variation as a function of time, increasing from an initial value of $\langle q_p \rangle_{zx}/q_i \approx 0.1$ and attaining a constant value of $\langle q_p \rangle_{zx}/q_i \approx 0.21$ during the final steady-state interval. This trend is related to the evolution of the mean bottom friction since the mean particle flowrate is directly proportional to the bottom shear stress. Similar as in the laminar cases, the extreme value of $\langle q_p \rangle_z$ continuously diverge from the mean value evidencing the increase in spatial variability of the particle flowrate as a function of the evolution of the bedforms.

As was mentioned in the introduction, it is customary to express the particle flow rate as a function of the Shields number Θ, and it has become a common practice in the sediment transport community to pursue scaling laws which relate the two quantities. Such scaling laws are especially relevant in engineering applications (cf. chapter 2). In order to analyze the particle flow rate obtained in the present work in light of such scaling laws, first $\langle q_p \rangle_{zx}$ is further averaged in time as $\langle q_p \rangle_{zxt}$. For the laminar cases, a single value of $\langle q_p \rangle_{zxt}$ is computed over the final part of the respective observation interval. For the turbulent case on the other hand, a time-varying quasi-instantaneous $\langle q_p \rangle_{zxt}$ is determined by computing the value in small time intervals of approximately 25 bulk time units (the same time intervals used to compute the friction velocity). Figure 6.5 shows the $\langle q_p \rangle_{zxt}$, normalized by the previously mentioned scales, as the function of the Shields number Θ, for all the cases. As was mentioned in chapter 2, in laminar flows, the Shields number is commonly computed by assuming a parabolic Poiseuille profile over the sediment bed, termed Θ_{Pois} (cf. equation (4.4) of chapter 4), and it has been shown that the normalized particle flowrate varies cubically as a function of Θ_{Pois} in the case of featureless bedload transport (cf. equation (4.10) and figure 4.5). For the purpose of direct comparison with the data of the featureless laminar bedload transport, $\langle q_p \rangle_{zxt}/q_v$ versus Θ_{Pois} of cases L1, L2 and L3 is plotted in figure 6.5(*a*). The corresponding data points of reference simulations and the fit

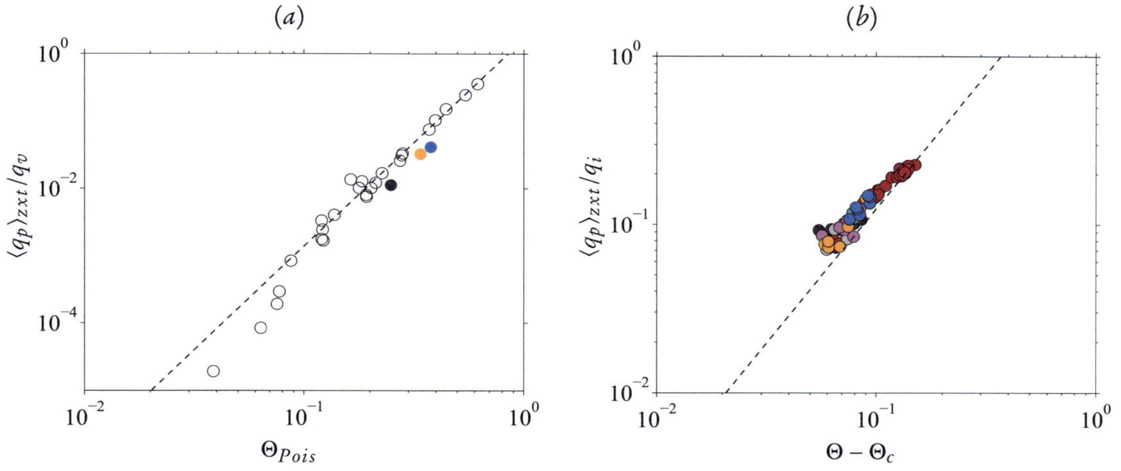

Figure 6.5: The space-time averaged volumetric particle flowrate $\langle q_p \rangle_{zxt}$ plotted versus the Shields number Θ. (*a*) For the laminar cases, ●, L1; ●, L2; ●, L3. The open circles are for featureless bedload transport in laminar flow (cf. figure 4.5 of chapter 4); the dashed line is the fit $\langle q_p \rangle_{zxt}/q_v = 1.66\,\Theta^{3.08}$ presented in equation (4.10) of chapter 4. Note that the Shields number computed under the assumption of a parabolic Poiseuille profile, Θ_{Pois}, is used in order to directly compare with the above mentioned scaling law. (*b*) For the turbulent case, ●, T1; ●, T1^{2H}; ●, T1^{3H}; ●, T1^{4H}; ●, T1^{6H}; ●, T1^{7H}; ●, T1^{48H}; . The dashed line is the Wong and Parker (2006) version $\langle q_p \rangle_{zxt}/q_i = 4.93\,(\Theta - 0.047)^{1.6}$ of the Meyer-Peter and Müller (1948) formula for turbulent flow. Note that all particle flowrate values are normalized by the viscous and inertial scales as in figure 6.4.

$\langle q_p \rangle_{zxt}/q_v = 1.66\,\Theta^{3.08}$ are also shown in the figure. It is seen that the present values are only slightly below the cubic power law fit and fall sufficiently within the scattered data, especially when the experimental data of Aussillous et al. (2013) is considered. The data points also suggest that the evolving bedforms have negligible influence on the scaling law as all points sufficiently fall on a cubic variation.

Turning to the turbulent case T1, figure 6.5(*b*) shows, that the quasi-instantaneous values $\langle q_p \rangle_{zxt}/q_i$, computed over the entire observation interval (except an initial small duration of approximately 50 bulk time units) plotted as a function of the non-dimensional excess bottom shear stress $\Theta - \Theta_c$ (note that in the turbulent case, the Shields number based upon the the actual shear stress, Θ, is used). A value of the critical Shields number $\Theta_c = 0.047$ is adopted (Meyer-Peter and Müller, 1948; Wong and Parker, 2006). To increase the sample confidence, corresponding data points of the other six independent turbulent simulations at identical parameter point as in case T1 but computed on different computational domain sizes are also presented (cf. table 5.1). Note that cases T1^{2H} and T1^{3H} have exhibited no evolution of patterns. Thus these cases correspond to a flow over a flat sediment bed. The figure also shows the empirical power law of Wong and Parker (2006), which is a modified version of the Meyer-Peter and Müller (1948) formula for turbulent flows,

$$\langle q_p \rangle_{zxt}/q_i = A(\Theta - \Theta_c)^{\alpha}\,, \tag{6.9}$$

where $A = 4.93$, $\alpha = 1.6$ and $\Theta_c = 0.047$. Note that Wong & Parker's formula is valid for plane sediment beds. As is observed in the figure, the DNS data points, which represent both plane sediment bed as well as pattern-featuring bed, are in a very good agreement with the scaling law (6.9). The fact that the present data agrees well with that prediction shows that the evolution of 'vortex dunes' does increase the net particle transport rate. However, the evolution also increases the net bottom shear stress. As a net result the formation of patterns does not strongly affect the particle transport scaling as a function of the excess shear stress (at least in the considered interval).

6.2 Dune-conditioned fluid flow and particle motion

In this section, we present an analysis of the turbulent flow field and particle motion which develops over the time-dependent sediment bed by focusing mainly on case T1. Let us start by looking at an instantaneous snapshot of the flow field as shown in figure 6.6. The flow field, which is taken at the end of the simulation interval of case T1, clearly shows the correlation between the spatially-varying sediment bed and that of the flow field. In figure 6.6(*a*), the intense coherent vortical structures are shown along with the position of the particles. It can be seen that the intensity and density of the coherent structures is the highest in regions downstream of the crest of the observed patterns. The contrary is observed for regions of the flow above and upstream of the crest where a lesser degree of turbulence activity is observed. The observed spatial distribution of vortices visually agrees with previous experimental and large-eddy simulation results which are largely based on the analysis of flow over fixed dunes McLean et al. (1994); Best (2005); Stoesser et al. (2008); Omidyeganeh and Piomelli (2011). The figure also suggests that the coherent structures leave their footprint in the bed shape, visible as slight longitudinal ridges and troughs superposed on the roughly two-dimensional dune patterns. This figure also illustrates the complexity of the flow structures in this configuration which, strictly speaking, does not possess any homogeneous spatial direction.

As was mentioned in chapter 2, the complex flow field over fixed or moving statistically two-dimensional dunes possesses several characteristic regions which can be summarized into an accelerating and a decelerating flow regions upstream and downstream of the dune crest respectively, a shear layer region downstream of the dune crest where flow separation occurs, a re-circulation region extending several dune heights downstream of the crest and bounded by the shear layer and a developing boundary layer region attached to the stoss side of the dune (see e.g. review by Best, 2005). Visual observation of the instantaneous field reveals some of these regions. Figures 6.6(*b*,*c*) show color contours of the instantaneous streamwise velocity field as well as the corresponding velocity fluctuations, with respect to the spanwise-averaged mean velocity, on a surface which is parallel to the spanwise-averaged sediment bed height but located at a wall-normal offset of one particle diameter above the fluid-bed interface (projected upon a horizontal plane). It is seen that the velocity field exhibits a clear spatial variation with largest values on the stoss side of the patterns reaching more than ten times the value of the friction velocity and smallest values in downstream side of the pattern crests. This highlights the accelerating/decelerating fluid regions which result from the form-induced spatial variation of the pressure gradient. The figure also shows the intersection of the surface with the re-circulation region characterized by the region where u_f attains negative-values. The spatial distribution of the streamwise velocity fluctuations additionally shows the distribution and length scales of the high-/low-speed fluid regions with respect to the position of the patterns. The spatial distribution of these velocity streaks is seen to be significantly different when compared to the high-/low-speed streaks of a single-phase flow over a smooth channel flow (see figure 10 of Kidanemariam et al., 2013, for instance). It is observed that the classical alternating streaks of high-/low-speed fluid regions are mainly observed in the developing boundary layer region. The velocity structures in the re-circulation region seem to be different in nature exhibiting no significant streamwise aligned coherence when compared to those located in the stoss side of the patterns.

The shear layer region is highlighted in figure 6.6(*d*) which shows the spanwise component of the instantaneous vorticity field on a streamwise/wall-normal plane located at $z = 0$ ($z = L_x$). As is expected, in the developing boundary layer region, i.e. on the stoss side of the patterns, the shear layer is mainly attached to the sediment bed. Starting from the crest of the dune however, the shear layer is observed to detach from the lee side of the patterns and it becomes more increasingly unstable and is seen to ultimately dissipate further downstream.

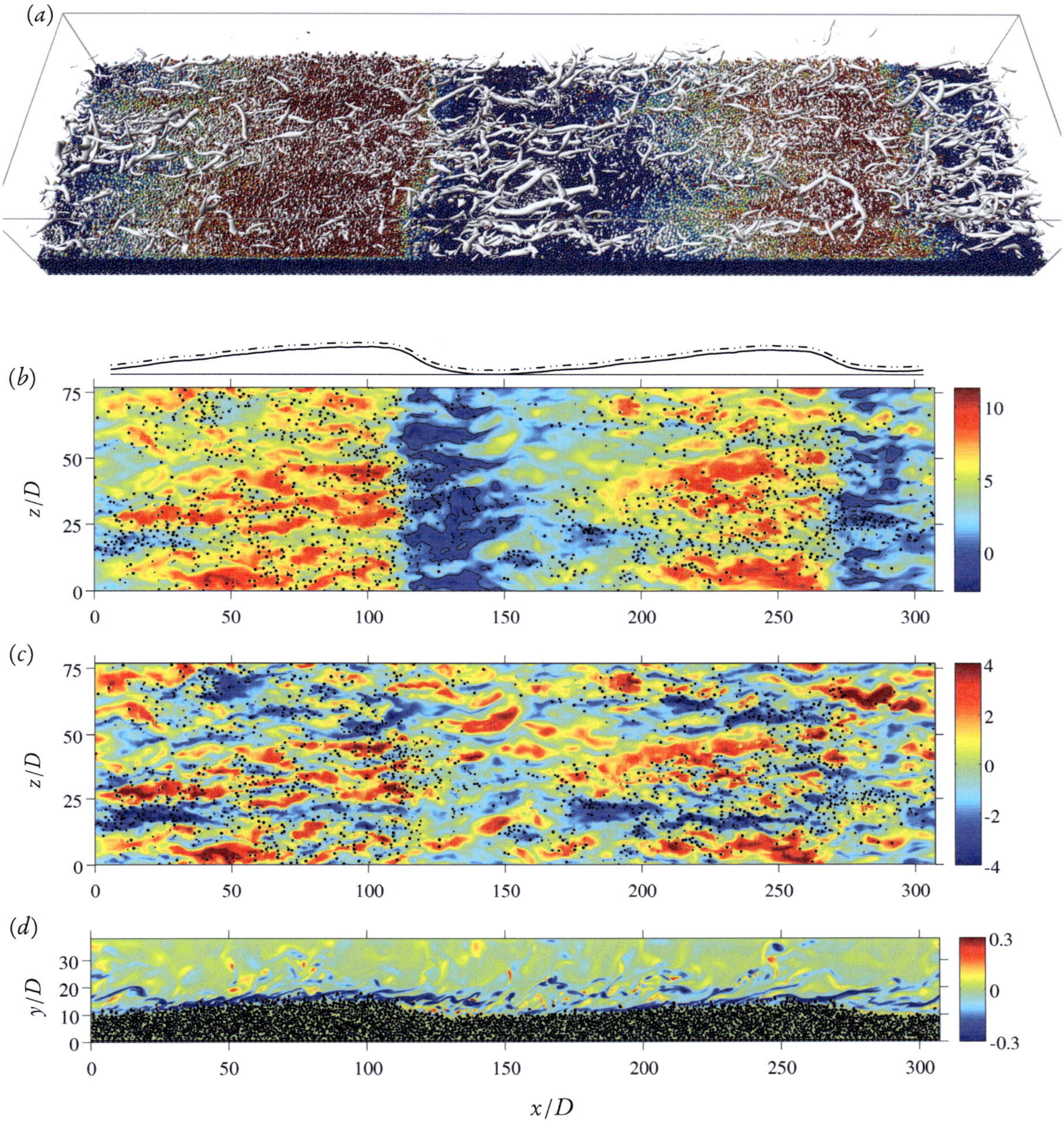

Figure 6.6: (*a*) Instantaneous three-dimensional snapshot of the flow field and particle positions of case T1 at the final time of the simulation interval ($tu_b/H \approx 1260$). Coherent vortical structures (grey surfaces) are educed by means of the λ_2 criterion of Jeong and Hussain (1995); particles are colored based on their wall-normal center location. (*b*) Color contour showing the streamwise component of the velocity field normalized by the friction velocity, u_f/u_τ, on a surface which is parallel to the fluid-bed interface but at a vertical offset of one particle diameter (shown by the dashed-dotted line shown at the top of the figure). Intersection region of particles and the surface is shown in black. The contour level corresponding to $u_f = 0$ is shown by th black lines. (*c*) Streamwise velocity fluctuations with respect to the spanwise-averaged instantaneous mean velocity field normalized by the friction velocity on the same surface. Intersection region of particles and the surface is shown in black. (*d*) Spanwise vorticity $\omega_z = \partial_x v_f - \partial_y u_f$, normalized by wall units, shown on an x–y plane located at $z = L_z$. Intersection region of particles and the plane is shown in black.

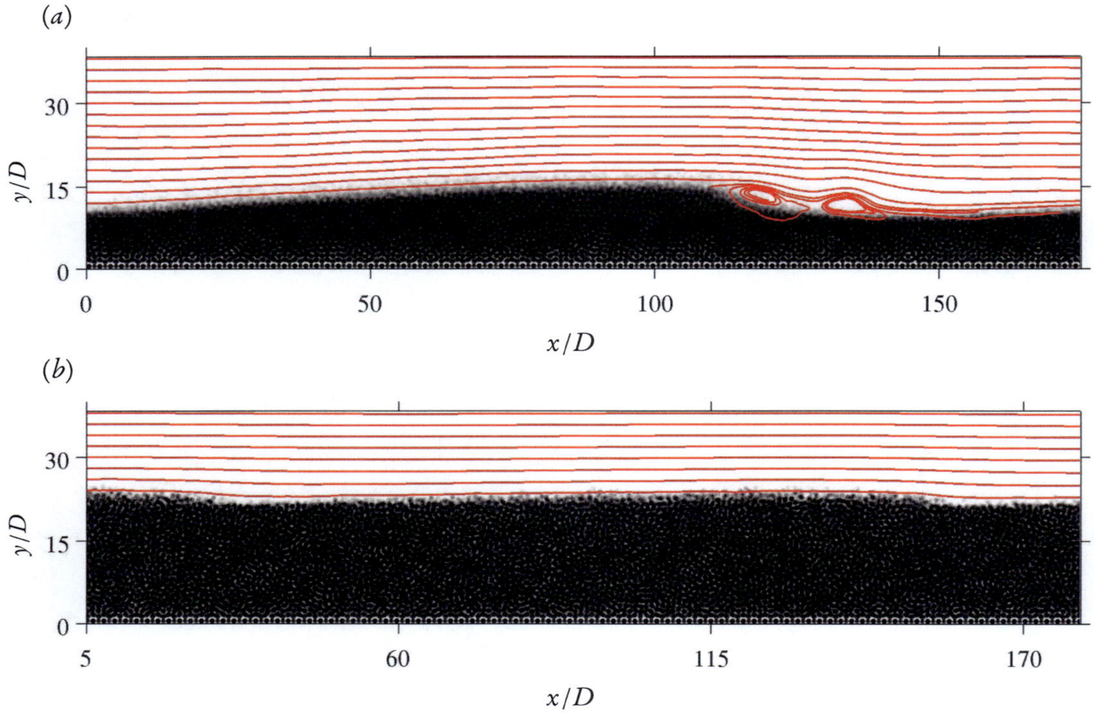

Figure 6.7: Close-up of the instantaneous, spanwise-averaged solid-volume fraction $\langle\phi_p\rangle_z$ (plotted in greyscale) as well as the streamlines computed from a spanwise-averaged instantaneous flow field at the final phase of the simulated interval: (*a*) case T1; (*b*) case L1.

It is believed that turbulence generated as a result of the instability of the shear layer which bounds the re-circulation zone is the dominant contributor to the turbulent kinetic energy budget of such type of flows (see e.g. Best, 2005, and references therein).

In order to compare the significance of the influence of the patterns on the spatial structure of the mean flow field between the turbulent and laminar cases, the streamlines of the spanwise averaged instantaneous flow field for the turbulent case T1 and the laminar case L2, corresponding to snapshots taken at the end of the simulated interval, is analyzed in figure 6.7. Consistent with results discussed in previous paragraphs, streamlines of case T1 are observed to be significantly influenced by the patterns, exhibiting flow separation on the lee-side of the patterns. The streamlines of the laminar case on the other hand are only slightly influenced by the presence of the patterns and no flow separation and re-circulation region is observed. The patterns in the laminar cases indeed correspond to 'small dunes' in the terminology of Ouriemi et al. (2009a), and to 'vortex dunes' those in the turbulent case. These results are, therefore, consistent with the regime classification based upon the channel (or pipe) Reynolds number proposed by these authors (cf. figure 5.1 in chapter 5).

Obviously, standard statistical tools, like those already performed in section 6.1, are not adequate to sufficiently characterize the structure of the turbulent flow such as that observed in figure 6.6. This inadequacy, which holds even for simpler configurations such as the flow over fixed two-dimensional smooth-surfaced dunes, has been already noted by several authors (see e.g. Yalin, 1977; McLean et al., 1994; Nikora et al., 2007a; Coleman and Nikora, 2011; Charru and Franklin, 2012) and various techniques have been proposed in the past, for instance the 'double-averaging' strategy advocated by Nikora et al. (2007a). Following the same spirit, we have performed 'dune-conditioned' spatial and temporal averaging of the flow field and particle motion, which takes into account the spatial and temporal variability of the sediment bed. The averaging procedure adopted in this study is similar to the procedure adopted by Charru and Franklin (2012). The main difference is that, Charru and Franklin (2012) consider a small averaging time interval such that the displacement of the dune was negligible during the observation, while the dune-conditioned averaging adopted in this study takes into account dune propagation. Moreover, in the present study, in addition to the fluid flow statistics, detailed dune-conditioned statistics of the particle motion is presented. To the best of our knowledge, no information on the motion of particles over an erodible, evolving sediment bed is available to the present date.

For the purpose of the averaging, a time interval has to be considered during which, the evolution of the patterns, i.e. the growth of their wavelength and amplitude is negligible. For case T1, the time interval in the range $t_s = 777 < tu_b/H < t_e = 1185$ covering approximately 410 bulk time units satisfies this condition. In this time interval, time evolution of the dimensions of the sediment bed exhibit not substantial change as a function of time (cf. figure 5.7). The propagation speed of the patterns is also approximately constant as can be seen in figure 5.9. Moreover, bulk fluid statistics, namely the box-averaged energy of the fluid fluctuations, the average bottom friction and the mean fluid velocity profile do not vary substantially with time (cf. figures 6.1, 6.2, and 6.3). The particle flowrate is also observed to be statistically constant in this interval (cf. figure 6.4). A total of 73 flow field and 2880 particle snapshots distributed evenly are used for the statistics.

6.2.1 Dune-conditioned averaging

Before presenting the results, let us give in this section a brief description of the dune-conditioned averaging applied in subsequent sections. For a given time t, within the above mentioned observation interval, a phase-shifted fluid velocity field $\widetilde{\boldsymbol{u}}_f$ is defined as

$$\widetilde{\boldsymbol{u}}_f(\boldsymbol{x},t) = \boldsymbol{u}_f(\boldsymbol{x} + \boldsymbol{\varphi}(t), t) \tag{6.10}$$

where the shift vector has only a streamwise component, i.e. $\boldsymbol{\varphi}(t) = (\varphi(t), 0, 0)^T$. For particle related Lagrangian quantities, for instance the particle velocity $\boldsymbol{u}_p$, first an instantaneous Eulerian field which has an information of the velocity of all particles, a "particle snapshot", is defined as

$$\boldsymbol{u}_p(\boldsymbol{x},t) = \sum_{l=1}^{N_p} \phi_p^{(l)}(\boldsymbol{x},t)\boldsymbol{u}_p^{(l)}(t) \tag{6.11}$$

where $\phi_p^{(l)}$ is the individual particle indicator function defined in appendix D. A similar phase-shifted particle velocity field is then defined as

$$\widetilde{\boldsymbol{u}}_p(\boldsymbol{x},t) = \boldsymbol{u}_p(\boldsymbol{x} + \boldsymbol{\varphi}(t), t) \; . \tag{6.12}$$

Other phase-shifted fluid and particles quantities are similarly defined. The amount of the phase shift φ is determined by the amount the patterns have propagated downstream from the reference time t_s until time t, which is computed from the shift of the maximum of the two-point/two-time correlation of the fluid-bed interface fluctuation $\langle h_b \rangle_z'(x,t)$ extracted at time t_s and t, viz.

$$R_{ht}(\delta x,t) = \langle \langle h_b \rangle_z'(x,t_s) \cdot \langle h_b \rangle_z'(x + \delta x,t) \rangle_x \tag{6.13a}$$

$$\varphi(t) = x \mid \forall\, \delta x \in [-L_x/2, L_x/2] : R_{ht}(\delta x,t) \le R_{ht}(x,t) \tag{6.13b}$$

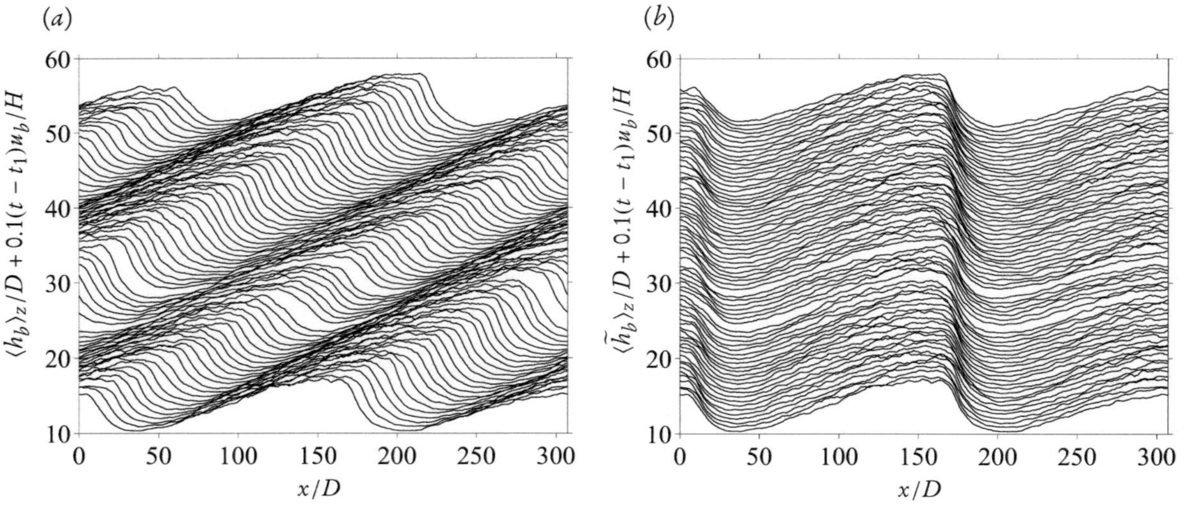

Figure 6.8: (*a*) Time-evolution of the sediment bed height, $\langle h_b \rangle_z$, in the steady-state interval The ordinate of the profiles is shifted vertically for clarity. (*b*) Same plot but for the phase-shifted sediment bed height $\langle \widetilde{h}_b \rangle_z$.

for all $t \in [t_s,\ t_e]$. Note that, due to the periodic boundary condition in the streamwise direction, all phase-shifted quantities circularly shift by an amount φ at the boundaries. It should also be noted that the above mentioned phase shifting procedure is related to but different from the procedure of observing the flow field from a frame of reference which is moving at a velocity which is equal to the propagation velocity of the patterns. In the latter, the amount of the propagation velocity has to be subtracted from the flow field. Figure 6.8 illustrates the phase-shifting procedure which shows the space-time plots of the actual sediment bed height $\langle h_b \rangle_z$, as well as the sediment bed height after applying the phase-shifting $\langle \widetilde{h}_b \rangle_z$, shown for the considered steady-state interval mentioned above.

Dune-conditioned statistics, which are functions of the streamwise and wall-normal directions, are computed by applying standard averaging of the phase-shifted fluid and particle data, in the spanwise-direction and over time (over the number of available fluid and particle snapshots). Furthermore, the instantaneous value of the phase-shifted fluid and particle fields are decomposed in the usual way into a mean and fluctuating components. For instance the phase-shifted fluid and particle velocity fields are decomposed as

$$\widetilde{\boldsymbol{u}}_f(\boldsymbol{x},t) = \langle \widetilde{\boldsymbol{u}}_f(\boldsymbol{x},t) \rangle_{zt} + \widetilde{\boldsymbol{u}}'_f(\boldsymbol{x},t) \qquad (6.14)$$

$$\widetilde{\boldsymbol{u}}_p(\boldsymbol{x},t) = \langle \widetilde{\boldsymbol{u}}_p(\boldsymbol{x},t) \rangle_{zt} + \widetilde{\boldsymbol{u}}'_p(\boldsymbol{x},t)\ . \qquad (6.15)$$

respectively. Covariances of the phase-shifted fluid and particle velocity fluctuations are defined similarly as $\langle \widetilde{\boldsymbol{u}}'_f(\boldsymbol{x},t) \otimes \widetilde{\boldsymbol{u}}'_f(\boldsymbol{x},t) \rangle_{zt}$ and $\langle \widetilde{\boldsymbol{u}}'_p(\boldsymbol{x},t) \otimes \widetilde{\boldsymbol{u}}'_p(\boldsymbol{x},t) \rangle_{zt}$ respectively. The averaging operator $\langle \cdot \rangle_{zt}$, which takes into account the fraction of space occupied by each of the phases, is precisely defined in appendix D.

It should be noted that, due to the streamwise periodicity, further streamwise averaging of the dune conditioned mean fluid and particle velocities results in the mean velocities of the respective phases which are obtained by performing standard averaging of channel flow. That is, $\langle \widetilde{\boldsymbol{u}}_f \rangle_{xzt} = \langle \boldsymbol{u}_f \rangle$ and $\langle \widetilde{\boldsymbol{u}}_p \rangle_{xzt} = \langle \boldsymbol{u}_p \rangle$. This is however not true for the velocity fluctuation covariances. For instance, considering the fluid phase,

$$\langle \boldsymbol{u}'_f \otimes \boldsymbol{u}'_f \rangle = \langle \widetilde{\boldsymbol{u}}'_f \otimes \widetilde{\boldsymbol{u}}'_f \rangle_{xzt} + \langle \widetilde{\boldsymbol{u}}''_f \otimes \widetilde{\boldsymbol{u}}''_f \rangle_{xzt} \qquad (6.16)$$

where $\langle \widetilde{\boldsymbol{u}}''_f \otimes \widetilde{\boldsymbol{u}}''_f \rangle_{xzt}$ is the dune-form induced velocity fluctuation covariance tensor which results from the deflection of the mean streamlines from the streamwise direction and is given by

$$\langle \widetilde{\boldsymbol{u}}''_f \otimes \widetilde{\boldsymbol{u}}''_f \rangle_{xzt} = \langle \langle \widetilde{\boldsymbol{u}}_f \rangle_{zt} \otimes \langle \widetilde{\boldsymbol{u}}_f \rangle_{zt} \rangle_x - \langle \boldsymbol{u}_f \rangle \otimes \langle \boldsymbol{u}_f \rangle \,. \tag{6.17}$$

The dune-conditioned velocity fluctuation covariances $\langle \widetilde{\boldsymbol{u}}'_f \otimes \widetilde{\boldsymbol{u}}'_f \rangle_{zt}$, on the other hand, correspond to the turbulent Reynolds stresses tensor. Such a decomposition of the covariance tensor into Reynolds stress tensor and dune-form induced stress tensor is similar to the 'double-averaging' concept of Nikora et al. (2007a). The only difference of the present averaging and theirs is the 'phase-shifting' of the velocity field.

6.2.2 Fluid and particle mean velocity

The streamwise and wall-normal components of the dune-conditioned fluid and particle velocities, $\langle \widetilde{u}_f \rangle_{zt}$, $\langle \widetilde{v}_f \rangle_{zt}$, $\langle \widetilde{u}_p \rangle_{zt}$ and $\langle \widetilde{v}_p \rangle_{zt}$ respectively, all normalized by the bulk velocity, are shown in figure 6.9. The figures show significant modulation of the averaged flow field which highlight the complex mutual interaction of the sediment bed and the driving flow. As is expected, below the fluid-bed interface, mean fluid and particle velocities are vanishing small. Above the fluid-bed interface however, velocities of both phases show large spatial variations which are seen to be strongly correlated with the dune geometry. At a given wall-normal offset from the dune surface, the value of the streamwise mean fluid velocity is observed to be higher in the region above the crest of the dune. It noticeably decreases downstream of the crest attaining negative values in the re-circulation region, which is shown in figure 6.9(*a*) by plotting a contour level corresponding to $\langle \widetilde{u}_f \rangle_{zt} = -0.05 u_\tau$. It should be noted, that a zero-valued contour level does not precisely demarcate the extents of the re-circulation region inside the sediment bed as there is non-zero mean velocity which corresponds to the very small value of the creeping flow velocity therein, as can be seen, for instance, in figure 4.8 in chapter 4. $\langle \widetilde{u}_f \rangle_{zt}$ gradually increases along the stoss side of the dune until it attains its maximum value in the region above the crest of the next downstream dune. The wall-normal component of the mean fluid velocity, shown in figure 6.9(*b*), is also observed to exhibit alternating upward moving fluid region ($\langle \widetilde{v}_f \rangle_{zt} > 0$) above the stoss side of the patterns and a downward moving fluid region ($\langle \widetilde{v}_f \rangle_{zt} < 0$) above the lee side of the patterns. A small region of positive $\langle \widetilde{v}_f \rangle_{zt}$ is observed downstream of the crest and in the vicinity of the fluid-bed interface which corresponds to the up-lifting fluid of the re-circulation region. Note that due to the free-slip boundary condition, $\langle \widetilde{v}_f \rangle_{zt} = 0$ at $y = L_y$. The spatial non-uniformity of the mean fluid velocity is further highlighted in Figure 6.9(*e*) which shows streamlines $\widetilde{\Psi}$ and equi-potential lines $\widetilde{\Phi}$ which satisfy

$$\langle \widetilde{u}_f \rangle_{zt}(x,y) = \partial_x \widetilde{\Phi}(x,y) = \partial_y \widetilde{\Psi}(x,y) \tag{6.18a}$$

$$\langle \widetilde{v}_f \rangle_{zt}(x,y) = \partial_y \widetilde{\Phi}(x,y) = -\partial_x \widetilde{\Psi}(x,y) \,. \tag{6.18b}$$

The acceleration and contraction of the fluid above the crest of the dune as well as the deceleration and expansion downstream of the crest is clearly visible. The re-circulation region which develops and extends several dune heights downstream of dune crest, is also clearly highlighted. The spatial variations of the streamwise and wall-normal components of the dune-conditioned particle velocity also exhibit similar correlation with respect to the dune geometry when compared to their fluid counterparts. From the intensity

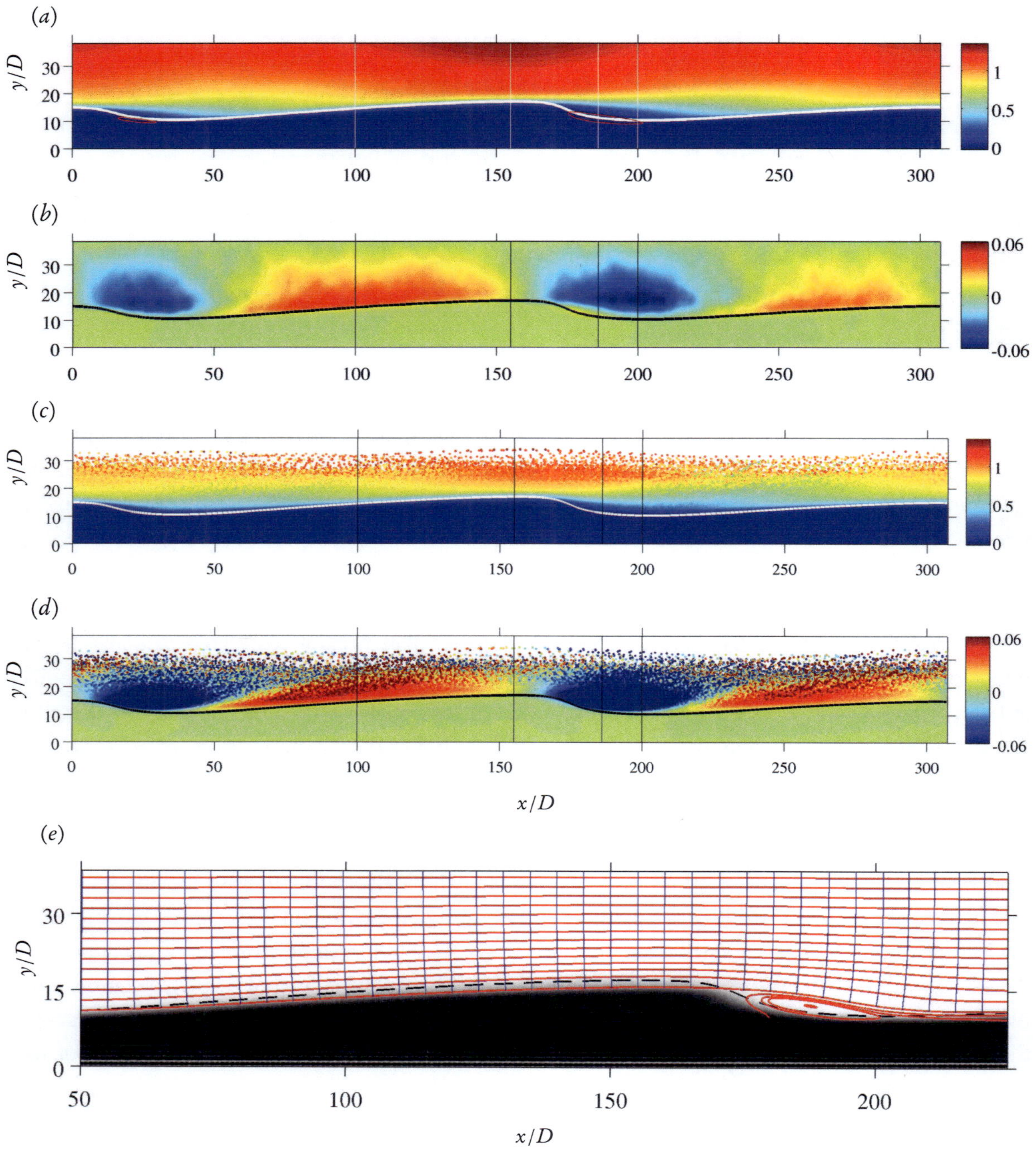

Figure 6.9: Dune-conditioned two-dimensional mean fluid and particle velocities. (*a*) Streamwise component of fluid velocity $\langle \widetilde{u}_f \rangle_{zt}/u_b$. Contour levels corresponding to a value of $\langle \widetilde{u}_f \rangle_{zt} \approx -0.05 u_\tau$ are marked by red lines. The mean of the phase-shifted sediment bed height, $\langle \widetilde{h}_b \rangle_{zt}$, is shown by the white thick line. The vertical thin lines indicate the locations where wall-normal profiles of each quantity are extracted and plotted in figure 6.11. (*b*) Wall-normal component of fluid velocity $\langle \widetilde{v}_f \rangle_{zt}/u_b$. (*c*) Streamwise component of particle velocity $\langle \widetilde{u}_p \rangle_{zt}/u_b$. (*d*) Wall-normal component of particle velocity $\langle \widetilde{v}_p \rangle_{zt}/u_b$. (*e*) Close-up of the spanwise-averaged dune-conditioned solid volume fraction $\langle \widetilde{\phi}_p \rangle_{zt}$ (plotted in greyscale) as well as the streamlines (red lines) and equi-potential lines (blue lines) computed from the dune-conditioned mean flow field. The extracted phase-shifted sediment bed is shown in dashed line.

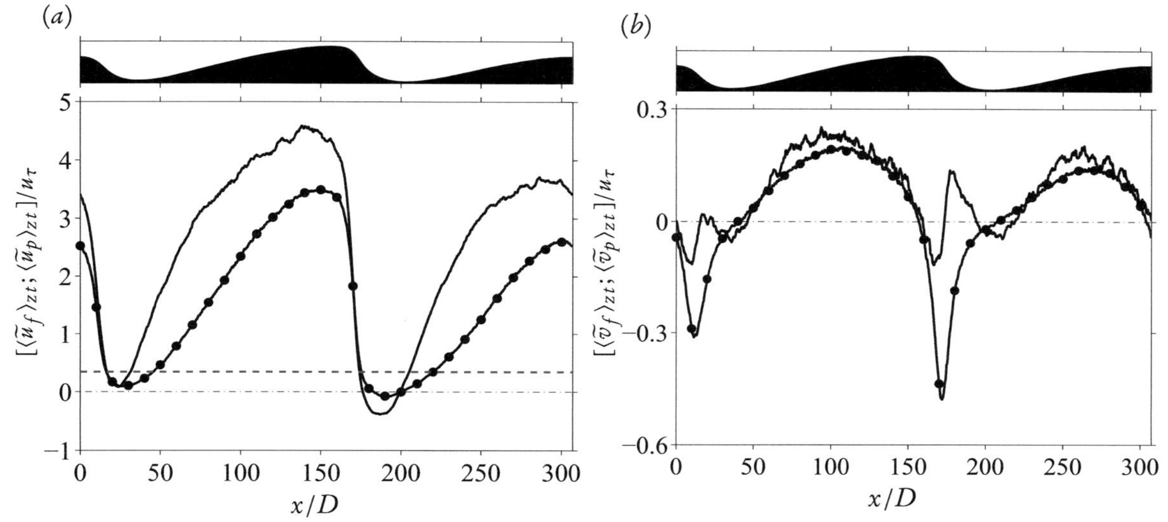

Figure 6.10: Dune-conditioned fluid velocity (—) and particle velocity (—•—) along the mean dune-conditioned sediment bed height $\langle \widetilde{h}_b \rangle_{zt}$ (shown by the black fill at the top of the figure). (*a*) Streamwise component of the velocity. The value of the mean propagation speed of the dunes in the considered observation time interval is shown by the shaded blue line (---) (*b*) Wall-normal component of the velocity.

of the color plots, it can be seen that the magnitude of streamwise particle velocity is on average smaller than that of the fluid in the fluid region above the fluid-bed interface, except in the re-circulation region. The opposite is true for the wall-normal component, i.e. the particle wall-normal velocity is observed to be higher in magnitude than that of the fluid.

More detailed quantitative comparison of the fluid and particle dune-conditioned mean velocities is provided in figures 6.10 and 6.11 which show profiles of the same quantities along the fluid-bed interface and along wall-normal lines located at selected streamwise locations (x/D = 100, 155, 186 and 200) respectively. Unlike most previous related studies, which consider flow over fixed dunes, velocities of both phases are not zero at the fluid-bed interface. This is due to the active exchange of mass and momentum (of both phases) which takes place at the fluid-bed interface as a result of erosion, transport and deposition of sediment particles as well as the fluid which flows in and out of the permeable moving surface. Moreover, figure 6.10 highlights the strong variation of velocity of both phases along the sediment bed height. At the dune crest streamwise velocity of both phases reach local maximum values of up to $\langle \widetilde{u}_f \rangle_{zt} \approx 4.5u_\tau$ and $\langle \widetilde{u}_p \rangle_{zt} \approx 3.5u_\tau$, whereas in the dune troughs, velocities decrease considerably and even attains negative values which result from the recirculating flow downstream of the dune crests. Another interesting observation is, that particles are on average moving at a smaller streamwise velocity than that of the fluid along the fluid-bed interface, except in the re-circulation region, where the opposite trend is observed. On the stoss side of the dunes, the value of the velocity lag between the two phases ranges between $0.75u_\tau$ and $1.75u_\tau$, where as in the re-circulation region, a negative velocity lag of up to $-0.35u_\tau$ is observed. The origin of the velocity lag could be explained by the combined effect of the preferential distribution of inertial particles with respect to the high- and low-speed fluid regions (Kidanemariam et al., 2013) and the fact that finite-sized particles, which are located at the interface, feel the traction effect of neighboring particles, which are located at a lower wall-normal location and are moving at a lower velocity, more than fluid particles do.

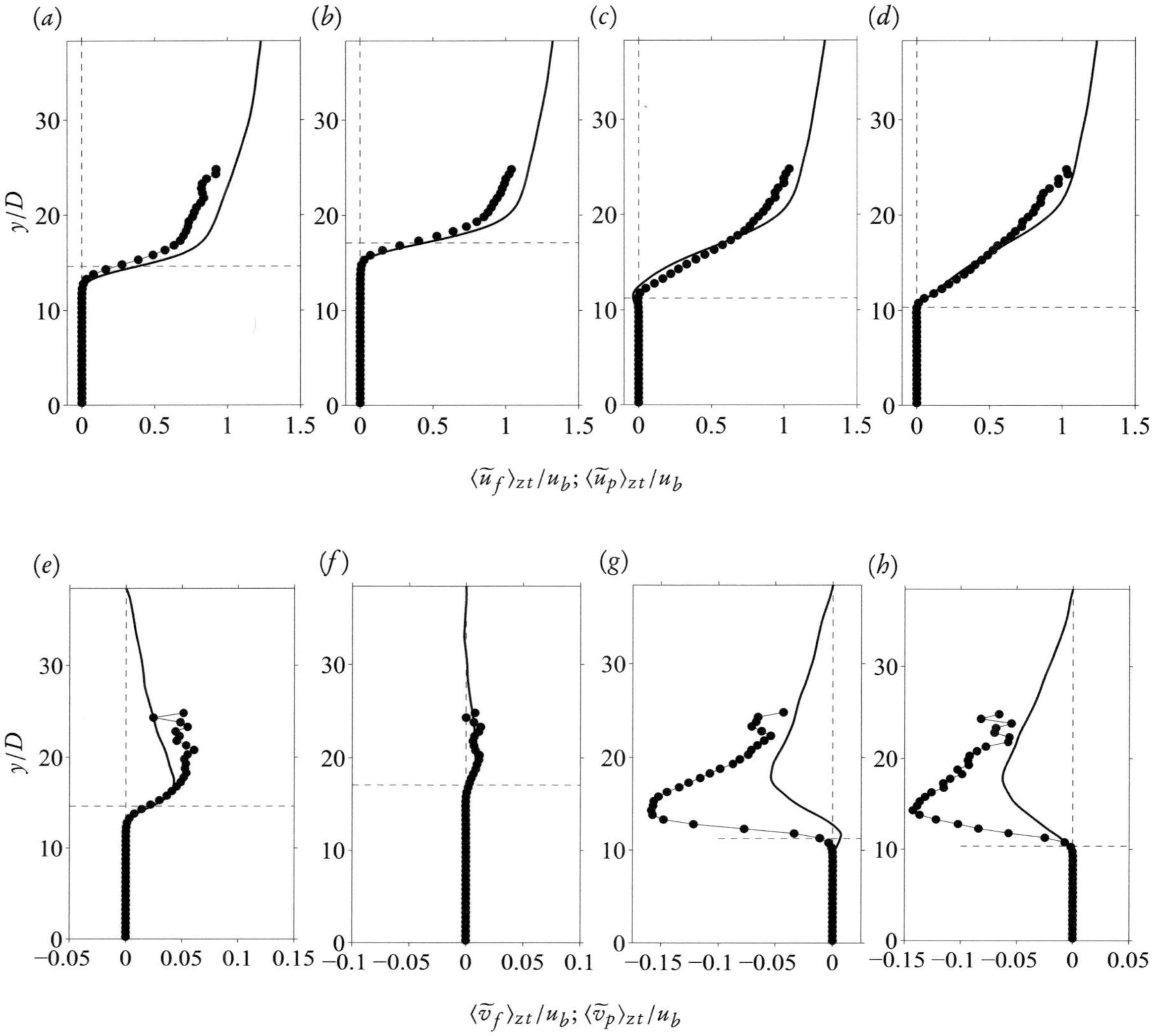

Figure 6.11: Wall-normal profiles of fluid and particle dune-conditioned mean velocities at selected streamwise locations. (*a–d*) $\langle\widetilde{u}_f\rangle_{zt}$ (—) and $\langle\widetilde{u}_p\rangle_{zt}$(—•—) at x/D = 100, 155, 186 and 200 consecutively. (*e–h*) $\langle\widetilde{v}_f\rangle_{zt}$ (—) and $\langle\widetilde{v}_p\rangle_{zt}$(—•—) at x/D = 100, 155, 186 and 200 consecutively. The horizontal dashed lines indicate the location of the fluid-bed interface at th corresponding streamwise location.

The wall-normal component of the fluid and particle velocity along the fluid-bed interface exhibit notably different trend when compared to their streamwise counterparts. In the stoss side, the difference between the mean wall-normal velocities of the phases is not as substantial as that of the streamwise component. Furthermore, the location of the local maximum of $\langle\widetilde{v}_f\rangle_{zt}(\langle\widetilde{v}_p\rangle_{zt})$ is observed to be upstream of the crest in contrast to that of the streamwise component. Both phases exhibit approximately zero value of mean wall-normal velocity at the location of the crest. The latter observation is a consequence of the dune-shape induced deflection of the mean flow from the horizontal as is highlighted by the streamlines in figure 6.9(*e*). Those streamlines are more deflected (higher wall-normal component) in the stoss side. On the other hand, the streamlines are approximately horizontally aligned (zero wall-normal component) at the location of the crest. Downstream of the crest, the particle velocity possess higher absolute value of the wall-normal component than that of the fluid. Comparing the mean streamwise velocity of sediments with the mean

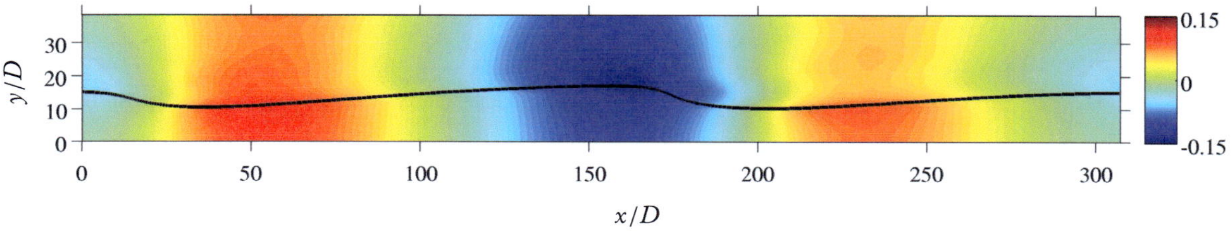

Figure 6.12: Dune-conditioned two-dimensional mean pressure $\langle\widetilde{p}\rangle_{zt}/(\rho_f u_b^2)$. The black thick line indicates the location of phase-shifted mean fluid-bed interface $\langle\widetilde{h}_b\rangle_{zt}$.

propagation velocity of the dunes, the former is much higher than the latter over the entire stoss side of the patterns. The opposite is true at locations downstream of the crest. This trend is in line with the fact that dunes propagate as a result of erosion and deposition of sediment grains.

The wall-normal profiles of fluid and particle velocities presented in figure 6.11 further corroborate the above observations. Note that the apparent velocity lag is significant even at wall-normal locations well above the fluid-bed interface. Namely, positive velocity lag in the stoss side of the dunes ($\langle\widetilde{u}_f\rangle_{zt} > \langle\widetilde{u}_p\rangle_{zt}$) and negative velocity lag in the re-circulation region ($\langle\widetilde{u}_f\rangle_{zt} < \langle\widetilde{u}_p\rangle_{zt}$). At wall-normal location well above the fluid-bed interface, the preferential sampling of low-speed fluid regions by inertial particles (Kidanemariam et al., 2013) is expected to be the main reason for the observed positive velocity lag. The negative velocity lag in the re-circulation region could be explained by the following argument. Sediment particles which are eroded from the stoss side of the dunes gain momentum as they get transported and approach the crest, attaining the highest momentum at the crest. They carry this high momentum into the re-circulation region and due to their inertia, do not immediately adapt to their new environment resulting on a statistically higher mean particle velocity than that of the fluid. Turning to the wall-normal velocity component, the profiles further highlight the mean positive and negative velocity upstream and downstream of dune crests respectively. Furthermore, upstream of the crest and above the fluid-bed interface, wall-normal velocity is observed to be larger in magnitude than that of the fluid where as downstream of the crest and above the fluid-bed interface, the particle velocity is observed to be substantially smaller (larger in absolute value) than that of the fluid. This latter effect is most probably due to the fact that gravity affecting the suspended grains, which settle in the recirculation region.

The overall trend of the variation of the mean streamwise and wall-normal components of the fluid velocity, with respect to the dune geometry, is consistent with the experimental observation made by Charru and Franklin (2012) where they have performed analogous analysis of the flow field over an isolated propagating three-dimensional barchan dune (measurements taken at the symmetry plane of the barchan). However, to our knowledge, there is no any available experimental data on particle velocity statistics over propagating dunes to compare to.

6.2.3 Mean pressure

Figure 6.12 shows the dune-conditioned mean pressure field, $\langle\widetilde{p}\rangle_{zt}$, normalized by $\rho_f u_b^2$. $\langle\widetilde{p}\rangle_{zt}$ exhibits a clear spatial variation (mainly in the streamwise direction) with alternating regions of positive and negative values. On average, a decrease in the value of the dune-conditioned mean pressure, i.e. a favorable pressure gradient, is observed in the region upstream of the dune crest and covering the stoss side of the dune. The opposite is true for the flow region downstream of the crest where an adverse pressure gradient is observed. This spatial variation in the pressure (and thus pressure-gradient) is driving the previously mentioned accelerating/decelerating fluid flow regions. Note that this 'dune-form-induced' spatially-varying pressure gradient is an additional forcing contribution to the imposed spatially-uniform time-varying pressure gradient which maintains a constant mass flow rate (cf. figure 6.2). Previous studies on flow over fixed two-dimensional dunes have also reported comparable observations (Stoesser et al., 2008; Omidyeganeh and Piomelli, 2011).

Unlike the velocity statistics, there exists a substantial spatial pressure variation inside the sediment bed. The sediment particles experience the resulting pressure gradient as part of the streamwise hydrodynamic force acting on their surface (in addition to the contributions from the imposed uniform pressure gradient). As has been shown in chapter 4, inside the sediment bed, the entire hydrodynamic forcing is counter balanced by the effective friction resistance between stationary particles (manifestation of the inter-particle tangential and normal collision forces). Note that, inside the sediment bed, since the relative velocity between the fluid and the particles is practically zero, there is no drag contribution to the total hydrodynamic force.

6.2.4 Covariances of fluid and particle velocity fluctuations

Figure 6.13 shows two-dimensional plots of the covariances of fluid and particle dune-conditioned velocity fluctuations, all normalized by u_τ^2. Additionally, figures 6.14 and 6.15 show streamwise and wall-normal profiles of the same quantities at selected locations for more quantitative observation. It should be noted that the dune-conditioned velocity fluctuations represent the turbulent fluctuations. The averaging procedure adopted here inherently excludes form-induced fluctuations which result from the spatial variation of the mean flow (cf. section 6.2.1). Furthermore, for the present case, the flow is statistically homogeneous in the spanwise direction. Thus, by definition, $\langle \widetilde{u}'_f \widetilde{w}'_f \rangle_{zt} = \langle \widetilde{v}'_f \widetilde{w}'_f \rangle_{zt} = \langle \widetilde{u}'_p \widetilde{w}'_p \rangle_{zt} = \langle \widetilde{v}'_p \widetilde{w}'_p \rangle_{zt} = 0$ if the convergence of statistics is guarantied.

The general spatial structure of the turbulent Reynolds normal stresses depicted in the figure 6.13(*a–c*), shows close similarity with that of the turbulence over fixed two-dimensional bedforms which have been extensively studied in the past (e.g. Engelund and Fredsoe, 1982; McLean and Smith, 1986; Nelson and Smith, 1989; Nelson et al., 1993; McLean et al., 1994; Bennett and Best, 1995; Stoesser et al., 2008; Omidyeganeh and Piomelli, 2011). Fluid velocity fluctuations, corresponding to all components, are observed to be maximum in the separated shear layer region downstream of the crest of the dunes, with values of the covariances reaching up to $6u_\tau^2$, $2u_\tau^2$, and $3u_\tau^2$ for the streamwise, wall-normal and spanwise components consecutively. This finding is in line with the general conclusion made by the above mentioned studies: most of the turbulent kinetic energy is produced in the separated shear layer region. As the shear layer expands further downstream (the wake region), the wall-normal and spanwise components of the velocity fluctuations are observed to decrease over the entire stoss side of the dune. $\langle \widetilde{u}'_f \widetilde{u}'_f \rangle_{zt}$ also decreases downstream. However, it attains a new maximum value within the developing internal boundary layer close to the fluid-bed interface. Thus, in the stoss side of the dunes, at a given streamwise location, $\langle \widetilde{u}'_f \widetilde{u}'_f \rangle_{zt}$ exhibits two local peaks, one corresponding to the developing boundary layer and another peak which corresponds to the expanding upstream shear layer, although the latter is much weaker in magnitude than the former. In figure 6.6(*c*), it is observable that the high- and low-speed streaks near the wall are more prominent when compared to the those in the recirculation region. This could be, among other factors, the reason for the increased value of near bed streamwise velocity fluctuations in the stoss side of the dunes.

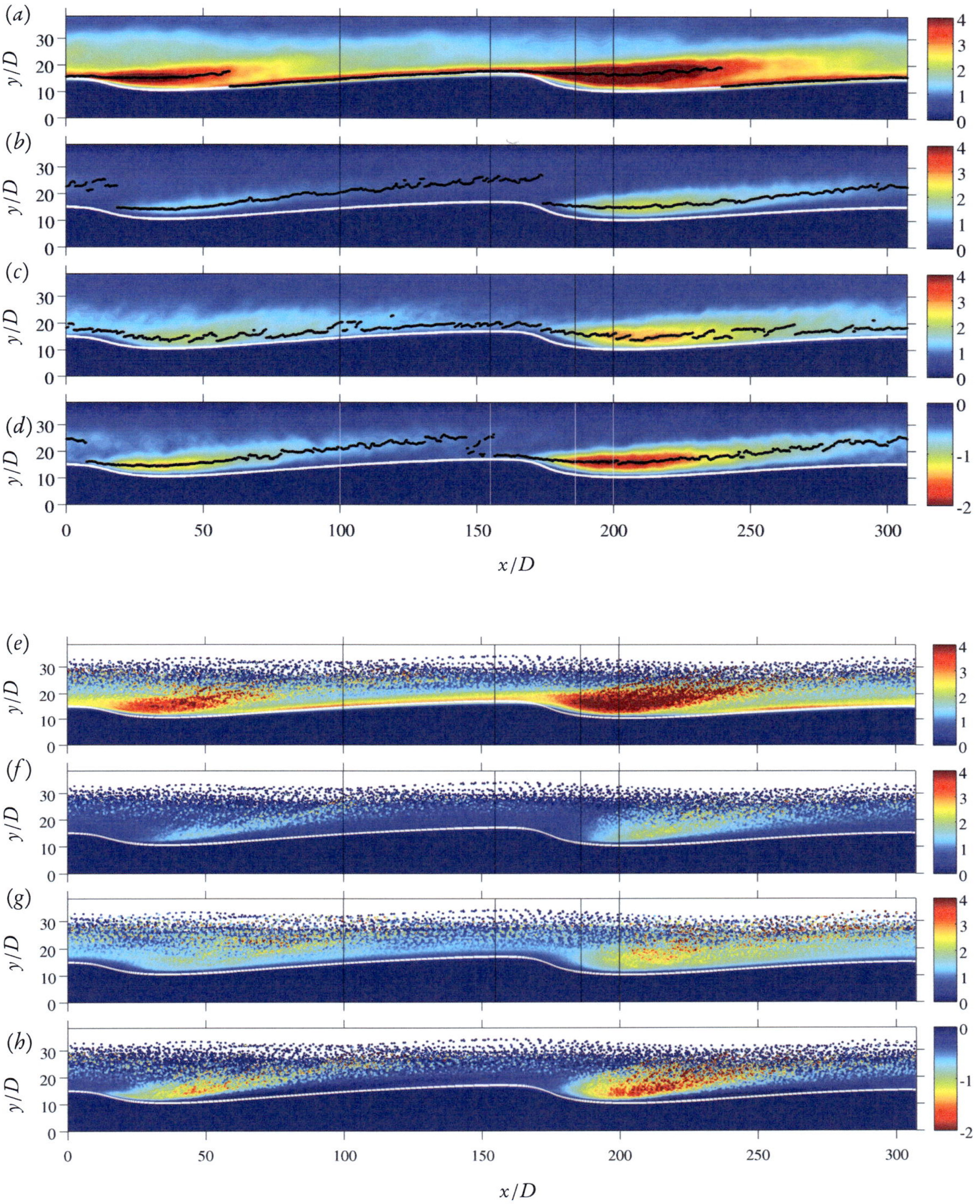

Figure 6.13: Covariances of dune-conditioned fluid and particle velocity fluctuations normalized by u_τ^2. (*a*) $\langle \widetilde{u}'_f \widetilde{u}'_f \rangle_{zt}$; (*b*) $\langle \widetilde{v}'_f \widetilde{v}'_f \rangle_{zt}$; (*c*) $\langle \widetilde{w}'_f \widetilde{w}'_f \rangle_{zt}$; (*d*) $\langle \widetilde{u}'_f \widetilde{v}'_f \rangle_{zt}$; (*d*) $\langle \widetilde{u}'_p \widetilde{u}'_p \rangle_{zt}$; (*e*) $\langle \widetilde{v}'_p \widetilde{v}'_p \rangle_{zt}$; (*f*) $\langle \widetilde{w}'_p \widetilde{w}'_p \rangle_{zt}$; (*g*) $\langle \widetilde{u}'_p \widetilde{v}'_p \rangle_{zt}$. In all plots, the white solid line indicates the location of the fluid-bed interface $\langle \widetilde{h}_b \rangle_{zt}$. The location of the maximum value of $\langle \widetilde{u}'_{f,\alpha} \widetilde{u}'_{f,\alpha} \rangle_{zt}$ at a given streamwise location is indicated by black dots in panels (*a–d*). The vertical lines indicate the location of sample wall-normal profiles shown in figure 6.15

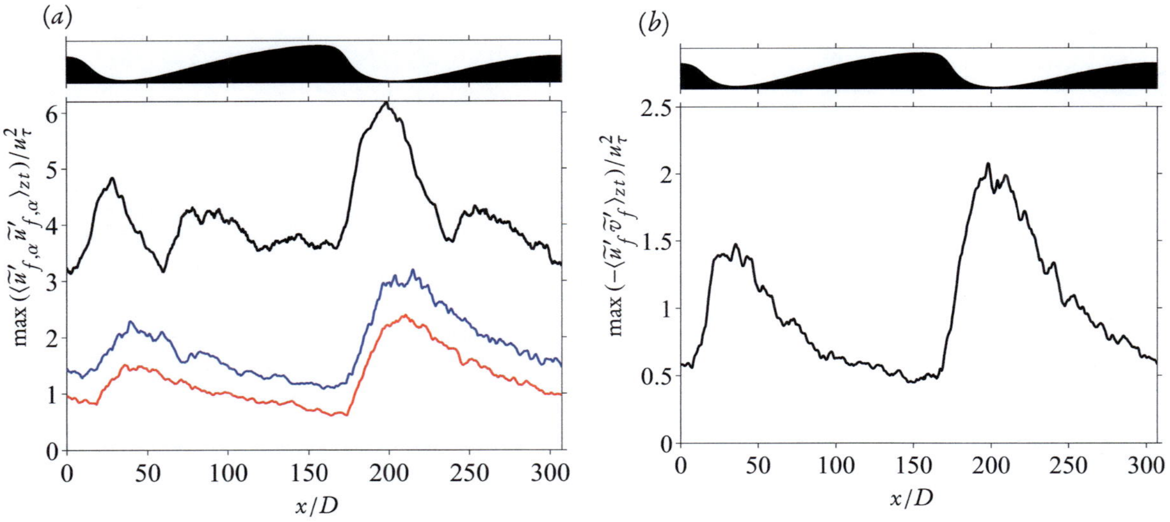

Figure 6.14: Streamwise variation of the peak value of the covariances of fluid velocity fluctuations at a given streamwise location. (*a*) $\langle \widetilde{u}'_{f,\alpha}\widetilde{u}'_{f,\alpha}\rangle_{zt}$. $\alpha = 1$,—; $\alpha = 2$,—; $\alpha = 3$,—. (*b*) $\langle \widetilde{u}'_{f}\widetilde{v}'_{f}\rangle_{zt}$. Note that these values are identical to those in figure 6.13(*a–d*) at the location of the black dots.

Furthermore, the streamwise component of the velocity fluctuations is observed to be larger in magnitude when compared to the other components showing a strong local anisotropic nature of the flow. Such anisotropy is, however, also a feature of standard channel flows. Nevertheless, the spatial variability of the degree of anisotropy and the streamwise and wall-normal non-homogeneity of the turbulence observed, is substantially different from that of a smooth-wall channel flow. On the other hand, in the outer flow region, far from the bottom, the turbulent statistics are comparable to those of a standard channel flow, although some differences are still observable. This corroborates the previous observations that, far away from the bottom, the influence of the bed undulations on the turbulent statistics is minor, at least for the present case. The similarity between the outer flow statistics and that of a standard channel flow has also been observed experimentally (see e.g. Cellino and Graf, 2000). The above described characteristics of the turbulence over the dunes are clearly observable from the sample wall-normal profiles of the Reynolds stresses at the selected streamwise locations provided in figure 6.15.

The dune-conditioned fluid Reynolds shear stress also shows similar spatial structure as the turbulent kinetic energy. $\langle \widetilde{u}'_{f}\widetilde{v}'_{f}\rangle_{zt}$ is observed to attain peak values in the separated shear layer regions. As is shown in figure 6.14(*b*), $\langle \widetilde{u}'_{f}\widetilde{v}'_{f}\rangle_{zt}$ attains values up to $2u_\tau^2$ downstream of the crests. It is then observed to monotonically decrease over the stoss side of the dune. As the crest is approached, the next downstream shear layer region becomes dominant and $\langle \widetilde{u}'_{f}\widetilde{v}'_{f}\rangle_{zt}$ increases again. Based on the knowledge accumulated from previous experimental, numerical as well as field observations, it is nowadays widely believed that the dune-related vortices, which are generated by Kelvin-Helmholtz type instabilities in the shear layer, are mainly responsible for the large values of Reynolds shear stress in this region (Best, 2005). The three-dimensional flow field shown in figure 6.6 also visually supports such claim, as the region downstream of the crests is seen to be more densely populated by strong coherent vortical structures.

In their recent contribution, Charru and Franklin (2012) compute the velocity fluctuations covariances of a turbulent flow over a single barchan dune, in a coordinate system which is local to the streamlines of the mean flow. Due to the small angle of deflection of the streamlines, streamwise and wall-normal components

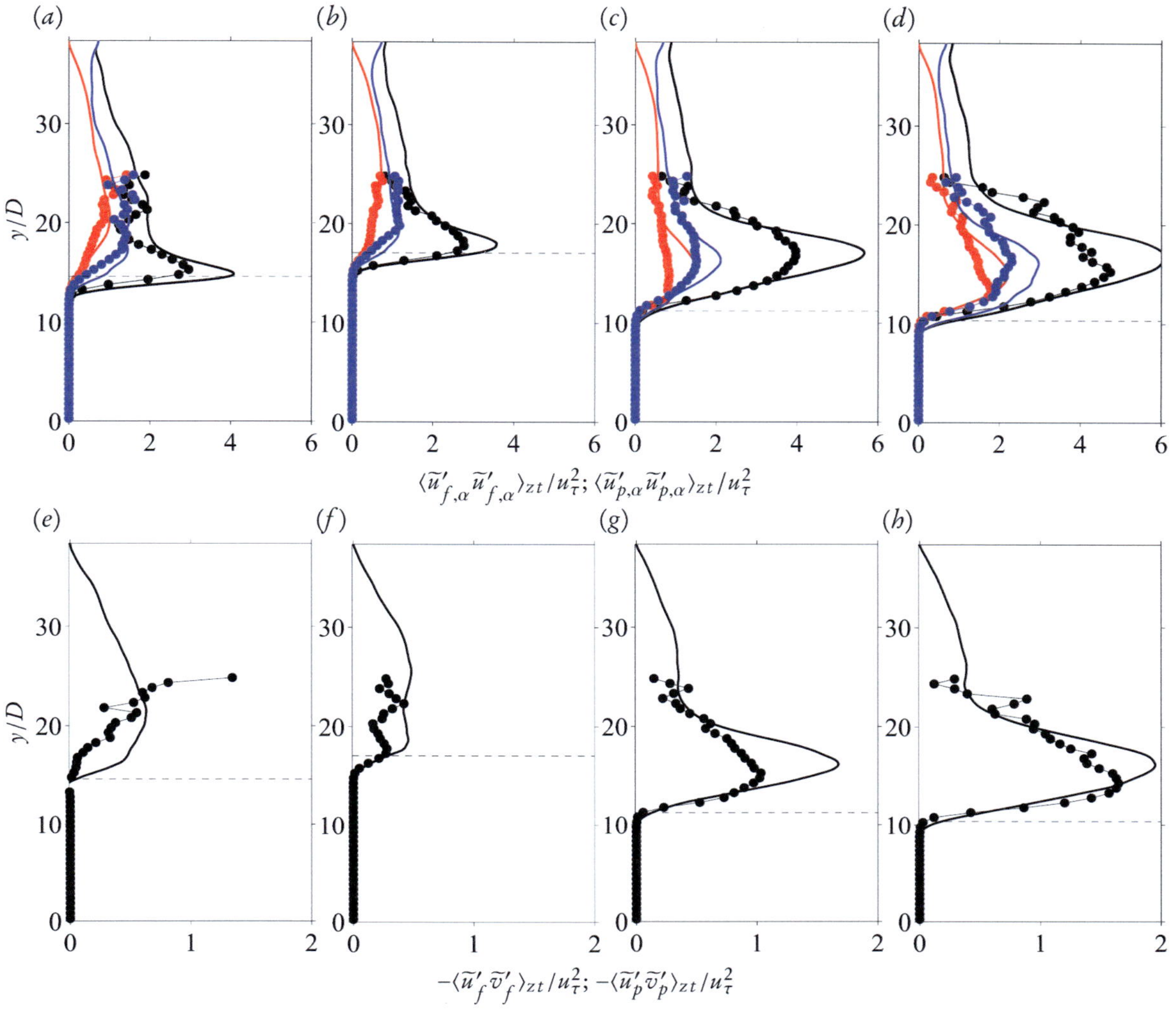

Figure 6.15: Wall-normal profiles of the covariances of fluid and particle dune-conditioned velocity fluctuations at selected streamwise locations. *(a–d)* $\langle \widetilde{u}'_{f,\alpha}\widetilde{u}'_{f,\alpha}\rangle_{zt}$ (—) and $\langle \widetilde{u}'_{p,\alpha}\widetilde{u}'_{p,\alpha}\rangle_{zt}$ (—•—) at x/D = 100, 155, 186 and 200 consecutively. The horizontal dashed lines indicate the location of the fluid-bed interface. Color coding similar as in figure 6.14. *(e–h)* Similar profiles but for $\langle \widetilde{u}'_{f}\widetilde{v}'_{f}\rangle_{zt}$ (—) and $\langle \widetilde{u}'_{p}\widetilde{v}'_{p}\rangle_{zt}$ (—•—).

of the velocity fluctuations in the Cartesian coordinate system can be safely compared to the tangential and orthogonal components respectively (Charru and Franklin, 2012). This assumption is not valid for the re-circulation region. However, Charru and Franklin (2012) did not report on the statistics in the recirculation region. Furthermore, Charru and Franklin (2012) consider a flow over a single barchan dune where there is no influence of an upstream located shear layer. Despite the differences, our results are comparable to theirs, especially with respect to the magnitude and location of the peaks of the Reynolds stresses.

Turning to the statistics of the particles, the dune-conditioned velocity fluctuations of the particles are observed to be strongly correlated to that of the fluid phase showing a similar trend. Nevertheless, some differences can be observed from the plots (cf. figures 6.13(*e–h*) and 6.15). Generally, intensities of particle velocity fluctuations and correlation between the streamwise and wall-normal particle velocity fluctuations are observed to be smaller and broader, than the that of the fluid phase. This observation is more enhanced for $\langle\widetilde{u}'_p\widetilde{u}'_p\rangle_{zt}$ which exhibits a damped peak both in the stoss and the lee side of the dunes. The wall-normal and spanwise particle velocity fluctuations are observed to be comparable to that of the fluid in the stoss side of the dunes. In the re-circulation region however, $\langle\widetilde{v}'_p\widetilde{v}'_p\rangle_{zt}$ and $\langle\widetilde{w}'_p\widetilde{w}'_p\rangle_{zt}$ exhibit smaller values than their fluid counterparts. A more isotropic trend of the particle velocity fluctuations has also been reported in particle-laden channel flow studies (Kidanemariam, 2010; Chan-Braun, 2012; Kidanemariam et al., 2013). It is believed that the action of inter-particle collisions, due to the relatively larger value of solid volume fraction in this region, tend to re-distribute the kinetic energy among the components. The cross-correlation coefficient $\langle\widetilde{u}'_p\widetilde{v}'_p\rangle_{zt}$similarly shows smaller values when compared to the fluid counterpart, over almost the entire reach of the dunes. It should be noted that, due to the scarcity of data samples in the outer flow region, no concrete observation can be made regarding particle statistics in this region.

6.2.5 Hydrodynamic and collision forces acting on particles

In this section we address the forces (hydrodynamic and inter-particle collision) acting on particles and their correlation with the geometry of the dunes. Figure 6.16 shows the spatial distribution of the dune-conditioned mean forces acting on particles, normalized by the submerged weight of a particle $F_W = (\rho_p - \rho_f)V_p|\mathbf{g}|$, decomposed into streamwise and wall-normal components.

Let us look first into the streamwise component of the forces. The hydrodynamic force $\langle\widetilde{F}^H_x\rangle_{zt}$ is observed to be positive in the stoss side of the dunes even deep inside the practically stationary sediment bed. It exhibits a clear increase in magnitude near the fluid-bed interface. Downstream of the crests, the sign of $\langle\widetilde{F}^H_x\rangle_{zt}$ is negative. Even those few particle which are suspended and are carried away by the mean flow, experience the alternating sign of the hydrodynamic force, experiencing positive drag in regions of the flow upstream of the crests and negative drag in the downstream side, with a clear minimum in the re-circulation region. Within the sediment bed and in regions of the fluid-bed interface, the collision force $\langle\widetilde{F}^C_x\rangle_{zt}$ exhibits a similar trend but of opposite sign when compared to the hydrodynamic counterpart. In regions of the flow above the sediment bed however, since the particle-particle collision is relatively infrequent, the collision forces are almost zero except for few occasionally colliding particle pairs. Regions of substantially large magnitudes of $\langle\widetilde{F}^C_x\rangle_{zt}$ (both positive and negative signs) are observed very close to the bottom wall of the computational domain. These alternating peaks correspond to the collision force acting on the stationary bottom particles which are used to mimic a rough bottom. As has been shown in section 6.1.2, it is mainly the sum of the collision force acting on these stationary particles which is balancing the imposed pressure gradient acting on the entire computational domain. As is shown in figure 6.16(*c*), the collision and hydrodynamic forces of the mobile particles balance each other and the total force (and thus particle acceleration) is almost zero inside the stationary sediment bed. At and above the fluid-bed interface however, as a result of the difference between the two forces, there are net particle acceleration and deceleration regions whose locations are strongly correlated with the dune geometry. The profile of the forces along the fluid-bed interface is explicitly shown in figure 6.17(*a*). An alternating trend of positive $\langle\widetilde{F}^H_x\rangle_{zt}$on the stoss side and negative $\langle\widetilde{F}^H_x\rangle_{zt}$ on the lee side is clearly observable. The streamwise variation of the forces is seen to be mild in the stoss side of the dunes with a more or less plateau like trend. The opposite trend is true for $\langle\widetilde{F}^C_x\rangle_{zt}$. Interestingly, the location of the maximum hydrodynamic force (the minimum of the collision force) is located downstream of the crests reaching values $\langle\widetilde{F}^H_x\rangle_{zt} \approx 0.65F_W$ and $\langle\widetilde{F}^C_x\rangle_{zt} \approx -0.6F_W$. This means that the hydrodynamic force acting on particles exhibits a phase shift between the dune geometry as well as with the mean flow velocity variation (as has been shown in figure 6.10). This finding could be linked to the conjectured phase shift between the bottom shear stress and dune geometry.

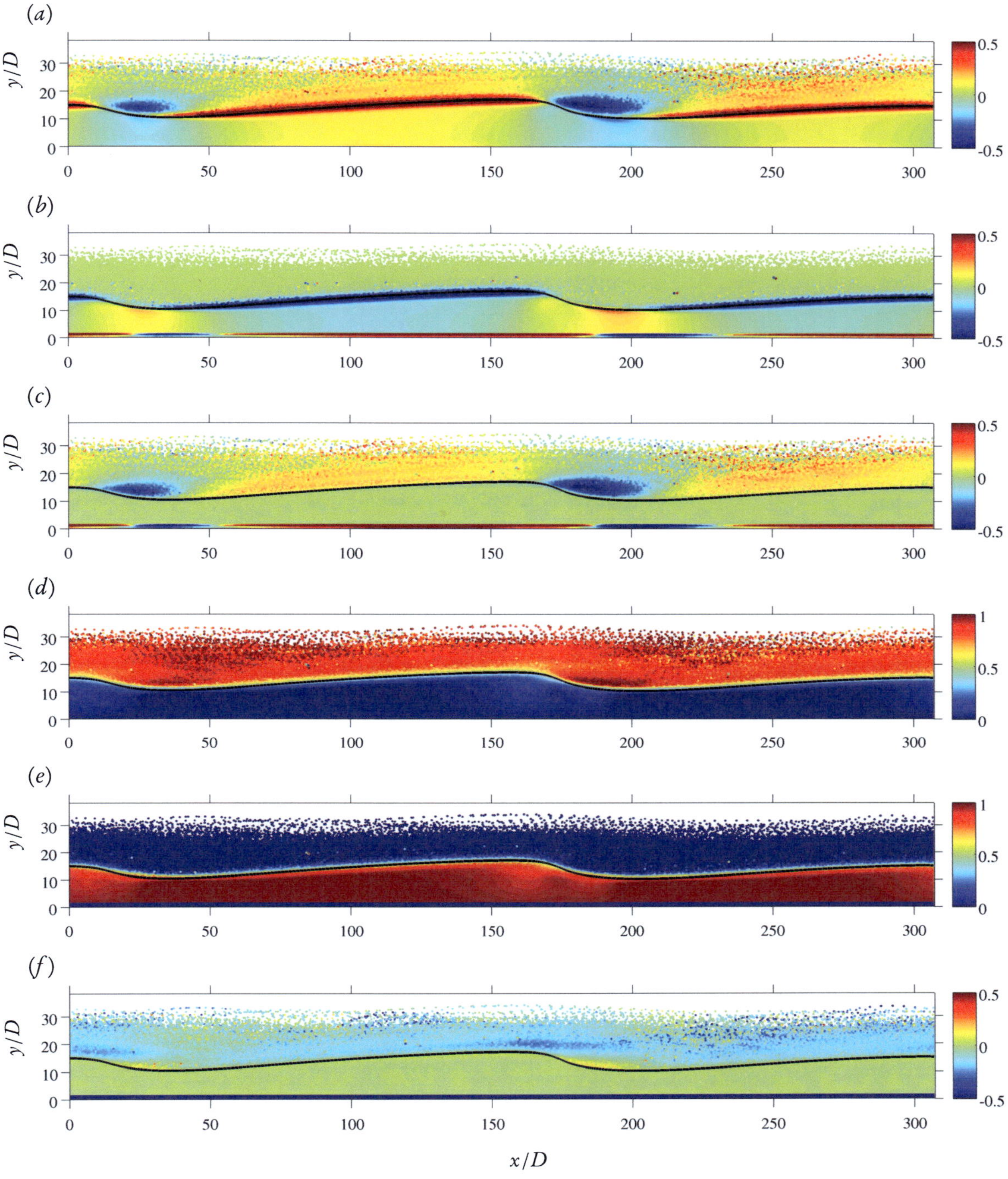

Figure 6.16: Dune-conditioned mean hydrodynamic and collision forces acting on particles. (*a*) Streamwise component of the hydrodynamic force $\langle \widetilde{F}_x^H \rangle_{zt}$. (*b*) Streamwise component of the collision force $\langle \widetilde{F}_x^C \rangle_{zt}$. (*c*) Total streamwise force acting on particles, $\langle \widetilde{F}_x^H \rangle_{zt} + \langle \widetilde{F}_x^C \rangle_{zt}$. (*d*) Wall-normal component of the hydrodynamic force $\langle \widetilde{F}_y^H \rangle_{zt}$. (*e*) Wall-normal component of the collision force $\langle \widetilde{F}_y^C \rangle_{zt}$. (*f*) Total wall-normal force acting on particles, $\langle \widetilde{F}_y^H \rangle_{zt} + \langle \widetilde{F}_y^C \rangle_{zt} - F_W$, where F_W is the submerged weight of a single particle, $F_W = (\rho_p - \rho_f) V_p |\mathbf{g}|$. In all plots, the forces are normalized by F_W.

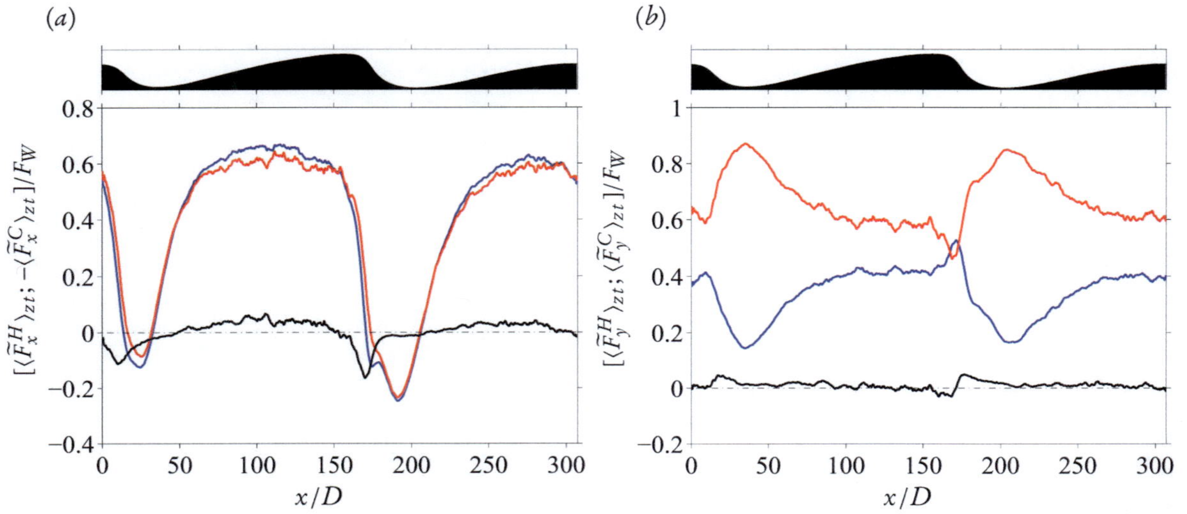

Figure 6.17: Streamwise variation of the dune-conditioned mean hydrodynamic force (—), collision force (—) and total force (—) acting on particles along the fluid-bed interface. (*a*) Streamwise component. (*b*) Wall-normal component. Note that the total wall-normal force is composed $\langle \widetilde{F}_y^H \rangle_{zt}$, $\langle \widetilde{F}_y^C \rangle_{zt}$ and F_W.

Let us turn to the wall-normal component of the forces. The total force acting on particles is the sum of hydrodynamic, collision and gravity forces. Figure 6.16 shows, that inside the stationary sediment bed, $\langle \widetilde{F}_y^H \rangle_{zt}$ is vary small when compared to $\langle \widetilde{F}_y^C \rangle_{zt}$. The collision force mainly balances the submerged weight of the particles and the net vertical force is zero. Far above the sediment bed, in the outer flow, the collision force acting on the few suspended particles is almost zero and, on average, particles experience an upward directed hydrodynamic force but with a magnitude which is slightly smaller than their submerged weight, $\langle \widetilde{F}_y^H \rangle_{zt}$ attaining maximum values between $0.85F_W$ and $0.9F_W$. Thus in the outer flow particles experience a net acceleration which is directed in the negative wall-normal direction. At and in the vicinity of the fluid-bed interface however, both forces are significant and contribute to the net force acting on the particles. As can be seen in figure 6.17(*b*), the sum of the hydrodynamic and collision forces is slightly larger than the submerged weight of the particles in most of the dune surface, especially noticeable in the recirculation regions.

6.2.6 Total shear stress

It has been previously shown, that the imposed mean pressure gradient in a statistically uni-directional particle-laden channel flow, is balanced by the sum of the fluid shear stress and stress contribution from the fluid-particle interaction (see e.g. Uhlmann, 2008; Kidanemariam et al., 2013). In the context of the immersed boundary method, the volume force, which imposes the no-slip condition at the fluid-particle interface, corresponds to the stress imposed on the system as a result of the fluid-solid interaction. When the equations of fluid motion are averaged over the entire domain comprising the fluid and particle domains, the streamwise momentum balance reduces to

$$-\left\langle \frac{\mathrm{d}p}{\mathrm{d}x} \right\rangle (L_y - y) = \underbrace{\rho_f \nu \partial_y \langle u \rangle - \rho_f \langle u'v' \rangle + \int_y^{L_y} \langle f_x \rangle \mathrm{d}y}_{\langle \tau_{tot} \rangle} \tag{6.19}$$

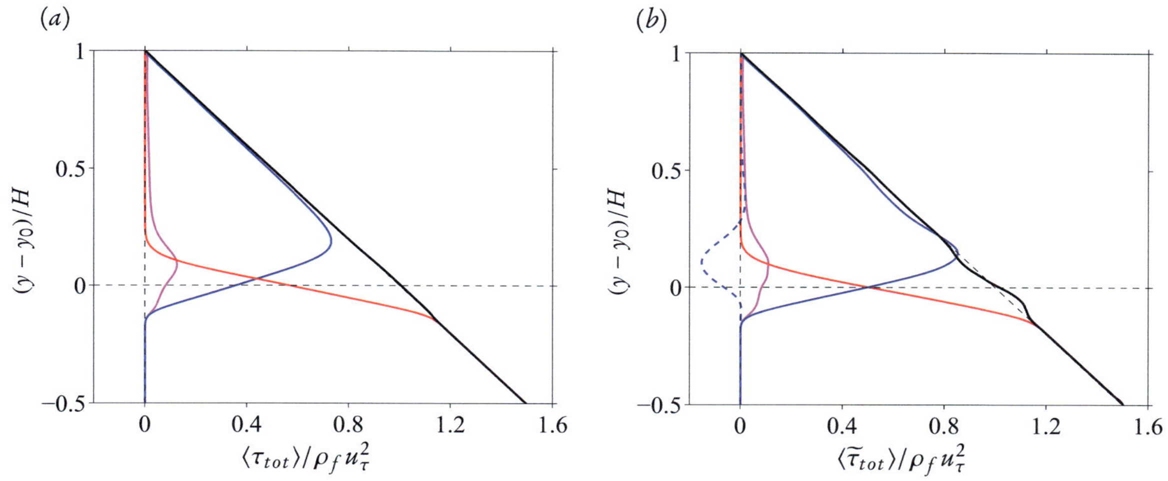

Figure 6.18: Wall-normal profiles of the different contributions to the space and time averaged total shear stress. (*a*) Total shear stress computed based on relation (6.19). —, $\rho_f \nu \partial_y \langle u \rangle$; —, $-\rho_f \langle u'v' \rangle$; —, $\int_y^{L_y} \langle f_x \rangle \mathrm{d}y$; —, $\langle \tau_{tot} \rangle$. (*b*) Total shear stress computed based on the dune-conditioned averaged statistics (relation (6.20)) —, $\rho_f \nu \partial_y \langle \widetilde{u}_f \rangle_{xzt}$; —, $-\rho_f \langle \widetilde{u}'_f \widetilde{v}'_f \rangle_{xzt}$; - - -, $-\rho_f \langle \widetilde{u}''_f \widetilde{v}''_f \rangle_{xzt}$; —, $-\int_y^{L_y} \langle \langle \widetilde{F}_x^H \rangle_{zt} \langle \widetilde{\phi}_p \rangle_{zt} \rangle_x \mathrm{d}y / V_p$; —, $\langle \widetilde{\tau}_{tot} \rangle$. $V_p \equiv \pi D^3/6$ is the volume of a single spherical particle.

where the last term in the RHS represents the fluid-solid interaction. When evaluating equation (6.19) for the present case, i.e. considering the dunes as roughness elements, the total shear stress, $\langle \tau_{tot} \rangle$, should vary linearly as a function of wall-normal distance, for the considered steady-state interval. This is confirmed in figure 6.18(*a*), which shows the wall-normal variation of the different contributions to the total shear stress. The figure highlights that, in the domain above the highest crest of the dunes, the fluid stress entirely balances the pressure gradient term, while in the domain below the lowest trough of the dunes, the particle-related forcing balances the driving force. In between these two regions, both contribution play a role.

The linear variation of the total shear stress can also be recovered from the dune-conditioned statistics. The total dune-conditioned fluid shear stress, after averaging in the streamwise direction, comprises three terms: the viscous stress term, the turbulent Reynolds stress term and the form-induced stress term (the 'dispersive' stress according to Nikora et al., 2007b). The dune-conditioned fluid-particle interaction forcing term is not explicitly available for evaluation. However, as has been shown by Uhlmann (2008), it can be approximated from the mean total hydrodynamic force acting on individual particles, i.e. $\langle f_x \rangle \approx -\langle F_x^H \rangle \langle \phi_s \rangle / V_p$. The error of this approximation is expected to be appreciable in locations where there is a large wall-normal gradient in $\langle \phi_s \rangle$. Thus the total shear stress should be sufficiently approximated by the following relation:

$$\langle \widetilde{\tau}_{tot} \rangle \approx \rho_f \nu \partial_y \langle \widetilde{u}_f \rangle_{xzt} - \rho_f \langle \widetilde{u}'_f \widetilde{v}'_f \rangle_{xzt} - \rho_f \langle \widetilde{u}''_f \widetilde{v}''_f \rangle_{xzt} - \frac{1}{V_p} \int_y^{L_y} \langle \langle \widetilde{F}_x^H \rangle_{zt} \langle \widetilde{\phi}_p \rangle_{zt} \rangle_x \mathrm{d}y \,. \quad (6.20)$$

Strictly speaking, the fluid velocity averaged over the composite domain should be used in relation (6.20). However, due to the relatively small difference between the mean streamwise velocity of both phases, the error of using only the velocity of the fluid, is expected to be small. Figure 6.18(*b*) shows the wall-normal profiles of the different contributions in (6.20). One can observe that the actual total shear stress is sufficiently recovered with this approach. A slight deviation from the linear variation is observable, in regions where the above introduced errors are expected to be substantial. Moreover, the figure shows, that the form-induced stress mainly works against the Reynolds stress, except for the fluid region above $y \approx 0.3H + y_0$ where it exhibits small positive values. This means that, in terms of standard averaging in wall-parallel planes, there is high momentum fluid which is apparently carried away from the bottom into the outer flow and vise versa as a result of the form-induced stress. The magnitude of the form-induced stress is however, small when compared to that of the Reynolds stress, attaining a maximum value which is approximately 0.18 times the maximum value of the Reynolds stress. The reversed effect of the apparent momentum transfer of the form-induced stress is easily explainable by looking into the streamline plot of the mean fluid velocity. In the stoss side of the dunes, the mean flow accelerates in the streamwise direction and is characterized by a positive wall-normal velocity (first quadrant stress), while on the lee side of the dunes, it decelerates and exhibits a negative value of wall-normal velocity (third quadrant stress).

6.2.7 Spatial variation of total boundary shear stress and particle flow rate

Total boundary shear stress

The analysis presented in section 6.2.6 provides an integral information on the total shear stress of the flow over a space and time varying sediment bed. However, it fails to provide information on the local boundary shear stress and its spatial variation with respect to the spatially varying bed. This quantity, and its relation with the local sediment flow rate, is the most relevant ingredient of all models of sedimentary patterns and thus its accurate determination is important. In developed bedforms, the total boundary shear stress is the sum of the dune-form drag, which results from the pressure difference between the upstream and downstream of the crest of a dune, as well as the skin friction, which amounts to the fluid viscous stresses as well as stresses as a result of fluid-particle interaction at the grain scale (see e.g. Yalin, 1977; Nikora et al., 2007b). Usually in experiments, the form drag is estimated by integration of pressure measurements along a stationary bedform and the skin friction is estimated by assuming a logarithmic velocity profile below the lowest velocity measurement point and performing extrapolation in which a certain value for the hydrodynamic roughness height has to be prescribed (see e.g. McLean and Smith, 1986; Maddux et al., 2003; Nikora et al., 2007b). Similarly, Charru and Franklin (2012) estimate the boundary stress by extrapolation of the total shear stress profile (corrected for streamline curvature) within the internal developing boundary layer to the surface of the bed. However, the degree of uncertainty inherent in such approaches is substantially large, and in the present case, it proved to be unreliable. Instead, we have determined the local boundary shear stress by evaluating momentum balance in a local control volume as described in the following paragraph. Note that determining the bottom friction based on the mean momentum balance is common in the boundary layer community (cf. 'von Kármán's integral momentum equation' in Pope, 2000, page 302).

The integral form of the Reynolds averaged momentum equation integrated over a constant shape control volume $\mathcal{V}$ which is enclosed by a surface $\mathcal{S}$ and which is moving at a constant velocity vector $\boldsymbol{u}_D$, can be written as

$$\frac{\partial}{\partial \widehat{t}} \int_{\mathcal{V}} \langle \widehat{\boldsymbol{u}} \rangle \mathrm{d}V + \oint_{\mathcal{S}} (\langle \widehat{\boldsymbol{u}} \rangle \otimes \langle \widehat{\boldsymbol{u}} \rangle) \cdot \widehat{\boldsymbol{n}} \mathrm{d}S + \oint_{\mathcal{S}} (\langle \widehat{\boldsymbol{u}}' \otimes \widehat{\boldsymbol{u}}' \rangle) \cdot \widehat{\boldsymbol{n}} \mathrm{d}S = - \oint_{\mathcal{S}} \langle \widehat{p} \rangle \overline{\overline{\boldsymbol{I}}} \cdot \widehat{\boldsymbol{n}} \mathrm{d}S + \oint_{\mathcal{S}} \langle \widehat{\overline{\overline{\boldsymbol{\tau}}}} \rangle \cdot \widehat{\boldsymbol{n}} \mathrm{d}S + \int_{\mathcal{V}} \langle \widehat{\boldsymbol{F}} \rangle \mathrm{d}V \quad (6.21)$$

where $\widehat{\boldsymbol{u}}$ is the velocity vector as observed from a moving inertial frame of reference (translating at the same velocity as the control volume). The derivation of equation (6.21) is provided in appendix C. In the present case, $\boldsymbol{u}_D$ corresponds to the mean dune propagation velocity in the positive streamwise direction. Moreover, since the dunes are migrating at a constant velocity $\boldsymbol{u}_D$, it follows that the shift vector $\varphi(t) = \boldsymbol{u}_D t$. Thus, using the definitions in (C.2), it can easily be shown that

$$\widetilde{\boldsymbol{u}}_f(\widehat{\boldsymbol{x}}, \widehat{t}) = \boldsymbol{u}_f(\boldsymbol{x}, t) = \widehat{\boldsymbol{u}}_f(\widehat{\boldsymbol{x}}, \widehat{t}) + \boldsymbol{u}_D \,. \tag{6.22}$$

Now lets consider a control volume $\mathcal{V}$ as depicted in figure 6.19 which is defined by the locations of the points A, B, C and D. The idea of the present analysis is to determine the total shear stress at the bottom (face AC) from the momentum balance in such a control volume, by evaluating each terms of equation 6.21 on each of the faces of $\mathcal{V}$ (as detailed in appendix C). In order to achieve this, the shape of the control volume is chosen in the following way. The top boundary (face BD) coincides with the top boundary of the computational domain while faces AB and CD are aligned perpendicular to the streamwise direction. The streamwise extent of the control volume is chosen such that the dune geometry is sufficiently resolved. Specifically, a value of $x_C - x_A \approx 2.5D$ is adopted, where x_A and x_C are the streamwise coordinates of faces AB and CD respectively. For the accurate determination of the boundary shear stress, the bottom face AC should be chosen carefully. If the location of face AC is either too high from the fluid-bed interface or too deep into the sediment bed, the resulting shear stress is not representative of the boundary shear stress along the surface dunes. Moreover, if a mean flow streamline starting from point A (as observed in the moving frame of reference), deviates from face AC substantially, then there would be a momentum flux through face AC as a result of the mean flow (cf. last terms of expressions(C.13) and (C.13)) and the computed shear stress would not be representative. Therefore, the first criterion we set for the choice of face AC is that it should be parallel to the streamline of the mean flow such that the momentum flux due to the mean flow across face AC is zero. Generally, face AC should be curved with the streamlines. Nevertheless, the curvature between x_A and x_B is small, and thus is ignored assuming a constant slope which is equal to $\tan\alpha$ (cf. the schematic diagram in figure 6.19). Of course, the streamlines in the recirculation region do not exhibit a well defined small curvature curves. Therefore, in the recirculation region, the above criterion does not hold.

Face AC could be chosen such that it coincides with the geometrically defined fluid-bed interface. However, the fluid-bed interface is generally not parallel to the streamlines of the mean flow (cf. figure 6.9(*e*)) and there is substantial momentum flux as a result of the mean flow at the interface. In practice, we have chosen the contour line defined by $\langle\widetilde{u}\rangle_{zt} = u_D$ (corresponding to $\langle\widehat{u}_f\rangle = 0$) as a bottom boundary of the control volume (face AC). Strictly speaking, the contour line $\langle\widetilde{u}\rangle_{zt} = u_D$ is not a streamline. However, the mean velocity across the contour is negligibly small (except in the recirculation region). Our choice of the location for face AC is physically motivated. The migration of dunes is as a result of erosion of sediment particles from the stoss side and deposition on the lee side. The erosion of sediments happens in a thin layer of the sediment surface, while the bulk sediment bed is stationary. Thus the sediment bed as a whole is not migrating (cf. online movies at http://dx.doi.org/10.1017/jfm.2014.284). Thus in the moving frame of reference, the stationary bed will be observed as moving in the upstream direction at a velocity equal to $-u_d$, while most of the eroded particles will be observed to move in the downstream direction at a velocity larger than zero, except those in the recirculation region (cf. figure 6.10). Thus, in the stoss side of the dunes, the sediment bed is sheared at a location which is, on average, moving in the positive streamwise direction at a velocity u_d and the contour corresponding to $\langle\widetilde{u}\rangle_{zt} = u_D$ sufficiently represents this 'bed shearing location'. On the other hand, in the recirculation region, the contour line does not represent the boundary shear location, rather the curve extends along the separated shear layer region as a result of the flow reversal. Therefore, in this region, the contour $\langle\widetilde{u}\rangle_{zt} = u_D$ is not representative of the location where the boundary shear stress should be computed. For this reason, in the recirculation region, face AC is made to coincide with the fluid bed interface so that the computed shear stress better represents the boundary shear stress. It should be mentioned,

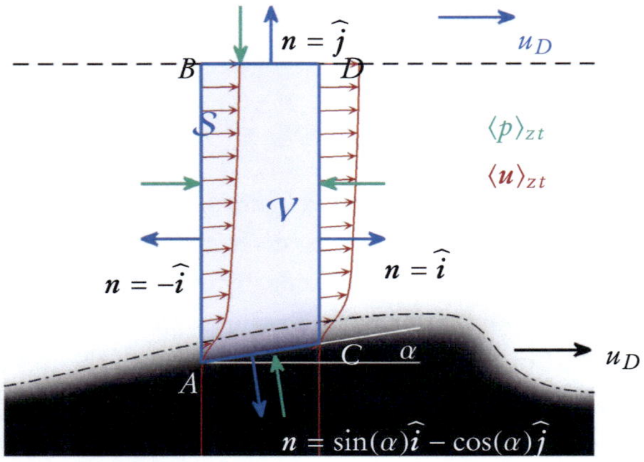

Figure 6.19: A Control volume over which the Reynolds averaged momentum equation are integrated in order to accurately determine the boundary shear stress. The control volume is moving at the constant mean dune velocity u_D in the positive streamwise direction. The streamwise extent of the control volume is chosen such that it is sufficiently smaller than the mean wavelength of the dunes, while the wall-normal extent is limited by wall-normal limit of the computational domain at the top and by the location where the mean fluid velocity in the moving frame of reference is zero, at the bottom.

that the precise determination of the local boundary shear stress along the stoss side of the dunes, where there is a net erosion of particles, is more important than in the recirculation region. It is well known that, as a result of the net deposition of particles in the recirculation region, the local particle flow rate (which is, on average, positive in this region) is not directly related to the local shear stress (which has a negative value).

The total resistive force acting on face AC has two contributions. The viscous resistive force as a result of velocity gradients as well as a shear force as a result of the fluid-solid interaction. Let us denote the total resistive force acting on $\mathcal{V}$ at face AC (per unit spanwise width) as $\boldsymbol{R} = R_x\widehat{\boldsymbol{i}} + R_y\widehat{\boldsymbol{j}}$. Then (cf. appendix C),

$$R_x = \widehat{f}_{visc,x} + \int_{\mathcal{A}} \langle \widehat{f}_x \rangle \mathrm{d}A\,, \quad \text{and} \quad R_y = \widehat{f}_{visc,y} + \int_{\mathcal{A}} \langle \widehat{f}_y \rangle \mathrm{d}A \tag{6.23}$$

where $\mathcal{A}$ is the x–y cross-sectional area of the control volume. The average total boundary shear stress (denoted as τ_{bot}) is then equal to but opposite in direction to the component of $\boldsymbol{R}$ along the direction AC, divided by the length of the segment AC viz.

$$\tau_{bot}(x_i) = -\frac{|\boldsymbol{R} - (\boldsymbol{R}\cdot\boldsymbol{n}_{AC})\boldsymbol{n}_{AC}|}{(x_C - x_A)}\cos\alpha \tag{6.24}$$

where x_i is a representative streamwise location of the control volume, say $x_i = (x_A + x_C)/2$. Note that the length of the segement $\overline{AC} = (x_C - x_A)\cos\alpha$. The fluid-solid interaction term in (6.23) implicitly accounts for the total stress as a result of all particles within the control volume, whether moving as bedload, suspended load or stationary.

The boundary shear stress can then be computed in two ways. Either by directly evaluating equation (6.23), or from the balance of momentum where each term of equation 6.21 is evaluated (as detailed in appendix C). In practice, specially in experimental studies, the fluid-solid interaction term is not easily accessible. Moreover, in the present analysis, although this term is accessible, we have not explicitly stored information of the Eulerian volume force field rather we have approximated it from the Lagrangian force acting on individual particles (cf.

section 6.2.6). It has been mentioned before that such an approximation introduces errors in regions where there is a sharp gradient in the solid volume fraction which is the case at the location of face AC. For this reason, and for demonstrating the determination of the boundary shear stress from experimentally accessible quantities, we have computed τ_{bot} by evaluating the momentum fluxes at all faces of $\mathcal{V}$.

Figure 6.20 shows the streamwise variation of the boundary shear stress τ_{bot}, computed with the above detailed method, along the dune surface. Note that, when computing the momentum flux due to pressure, the two contributions to $\langle p\rangle_{zt}$ have been considered, namely the contribution from the imposed pressure gradient as well as that from the form induced pressure gradient, i.e.

$$\langle p\rangle_{zt}(x,y) = \langle \mathrm{d}p/\mathrm{d}x\rangle x + \langle \widetilde{p}\rangle_{zt}(x,y) + p_0 \ . \tag{6.25}$$

It is seen than the bottom shear stress is the highest upstream of the dune crests, exhibiting a somewhat plateau-like region in the stoss side of the dunes. The value of τ_{bot} attained in the stoss side ranges between $1.5\rho_f u_\tau^2$ to $1.8\rho_f u_\tau^2$. These range of values are comparable with those reported by Charru and Franklin (2012) for the flow over barchan dunes. On the other hand, Charru and Franklin (2012) report that the maximum of the shear stress to be located at the crest of the barchan dune in contrast to the findings of the present study. Downstream of the crests, the shear stress attains negative values, which is expected due to the reverse flow of the recirculation region. The observed positive phase lag between the local boundary shear stress and the sediment bed height is believed to be an important factor for destabilizing a mobile sediment bed during the initial stages of instability (Charru et al., 2013; Andreotti and Claudin, 2013). As is clearly seen from the figure, not only is there a phase lag between τ_{bot} and $\langle \widetilde{h}_b\rangle_{zt}$, but also a difference in the shape of the variation. Nevertheless an average phase shift could be estimated from the location of the maximum of the cross-correlation between the two quantities as follows:

$$\begin{aligned} R_{h\tau}(\delta x) &= \langle \langle \widetilde{h}_b\rangle'_{zt}(x) \cdot \tau'_{bot}(x+\delta x)\rangle_x & (6.26a)\\ \varphi_{h\tau} &= x \mid \forall\, \delta x \in [-L_x/2, L_x/2] : R_{h\tau}(\delta x) \le R_{h\tau}(x)\ . & (6.26b) \end{aligned}$$

It turns out, that the average phase shift between τ_{bot} and $\langle \widetilde{h}_b\rangle_{zt}$ is $\varphi_{h\tau} \approx 15D$.

Volumetric particle flow rate

Figure 6.20 also shows the dune-conditioned mean local particle flow rate $\langle \widetilde{q}_p\rangle_{zt}$. This quantity is accurately computed by using equation (6.5) and applying the phase-averaging procedure. It is seen, that the particle flow rate is in phase with the dune geometry, with maximum value at the crests and minimum value at the troughs. Such an in-phase variation of the bed and particle flow rate is a consequence of the conservation of mass of the sediment bed, as described by the Exner equation (2.20) (Charru et al., 2013). Consequently, the average phase shift between the local boundary shear stress and the particle flow rate is approximately $15D$. As was discussed in the introduction, most earlier theoretical studies assume that the local particle flowrate is a function of the local shear stress, i.e. the shear stress and particle flow rate are in phase. Recent studies, on the other hand, incorporate the phase lag between τ_{bot} and $\langle \widetilde{q}_p\rangle_{zt}$ into their stability analysis (see e.g. Charru, 2006; Fourrière et al., 2010). In these studies, the effect of such a phase lag, which is termed as the 'saturation length scale', L_{sat}, is assumed to act as a stabilizing factor, while the phase lag between shear stress and topology acts as a bed destabilizing factor. Fourrière et al. (2010) and Claudin et al. (2011) mention that, for bedload transport, the relaxation length is of the order of 10 particle diameters and that the initial wavelength of subaqueous ripples is approximately $10L_{sat}$. The values of L_{sat} and λ_c/L_{sat} obtained in our DNS data support these predictions.

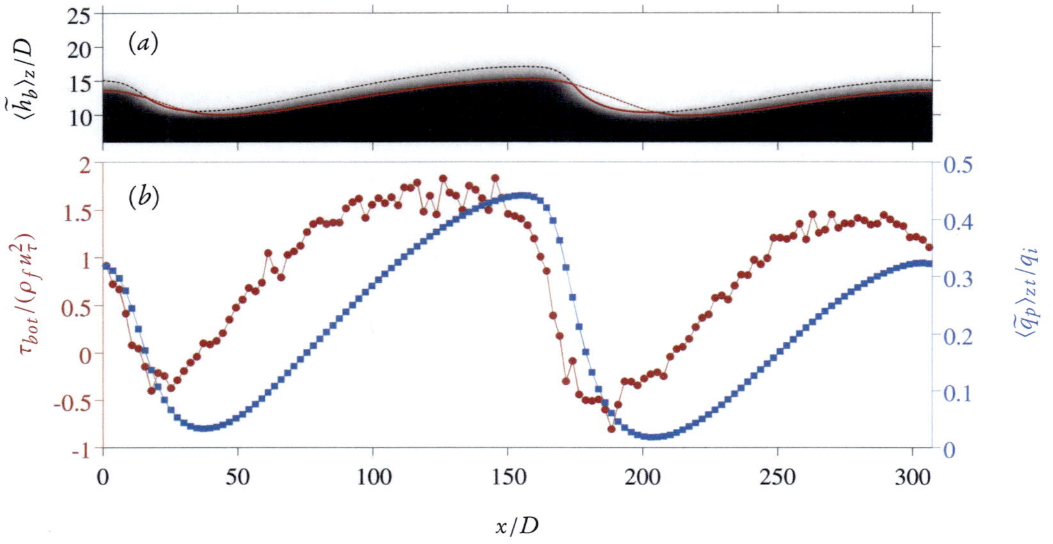

Figure 6.20: (*a*) The spanwise-averaged dune-conditioned solid volume fraction $\langle\widetilde{\phi}_p\rangle_{zt}$ (plotted in graysclale). The sediment bed height is shown in black dash-dotted curve. The red curve in the top panel corresponds to a contour level of $\langle\widetilde{u}_f\rangle_{zt} = u_D$ which is defined to be the bottom extent of the averaging volume shown in figure 6.19. Note that in the recirculation region, the contour $\langle\widetilde{u}_f\rangle_{zt} = u_D$ is located above the fluid bed interface (red dashed line). Thus in this region the fluid-bed interface is chosen as a bottom boundary of the control volume. (*b*) Streamwise variation of dune-conditioned boundary shear stress τ_{bot}, normalized by the average shear stress ρu_τ^2, (—•—) and the dune-conditioned volumetric particle flowrate $\langle\widetilde{q}_p\rangle_{zt}$, normalized by the inertial scale $q_i = U_g D$ (—▪—).

Nevertheless, it should be noted, that the computed phase shift between τ_{bot} and $\langle\widetilde{q}_p\rangle_{zt}$ corresponds to the dunes which are sufficiently beyond their initial formation stage. As was mentioned by Charru et al. (2013), it is conjectured that the phase shift decreases with time as the dunes evolve. Thus the value reported here might not represent the value of the phase shift during the instants of the instability. The phase lag during the initiation and evolution of the bedforms could be accurately determined applying the same momentum balance procedure applied to the DNS data of case T1$^{48\mathrm{H}}$ (which accommodated eleven initial wavelengths) in small time intervals during and after the formation of the bedforms. That case also allows further averaging over the number of dunes available in a given instant, thereby increasing the data sample. This task is planned for future work. Moreover, as was mentioned previously, although there are regions where the boundary shear stress is negative, the particle flow rate is observed to be above zero across the whole streamwise extent. Previous experimental studies have also reported similar observation of non-zero particle flow rate in the recirculation region (see e.g. McLean et al., 1994). This shows, that the spatial structure of sediment transport can not be entirely predicted by classical models where particle flow rate is expressed in terms of the bottom shear stress alone. This aspect deserves further investigation.

6.3 Summary and conclusion

This chapter was devoted to the detailed analysis of the flow field, as well as the associated particle motion, over the subaqueous sediment patterns in both the laminar and turbulent cases. First, the effect of evolution of the sediment bed on the bulk characteristics of the flow is investigated in detail. More specifically, the time evolution of the box-averaged fluid velocity fluctuations and mean bottom shear stress is studied. It is found that the evolution of the 'small dune' patterns has almost no influence on the bulk statistics of the driving laminar flow. This is confirmed by scrutinizing the time variation of the mean velocity profiles, shear stress profiles as well as the driving imposed pressure gradient (which is determined *a posteriori* from the balance of driving and resisting forces). On the contrary, the 'vortex dunes' which evolved in the turbulent case, are observed to substantially alter the bulk features of the flow. Analysis of the box-averaged fluid velocity fluctuations revealed, that the energy contained in the fluctuations of the turbulent case increases by more than twice the initial value, during the simulated observation interval of approximately 1200 bulk time units. The initial value is comparable to that of a smooth wall channel flow. Similarly, the mean bottom friction of the flow is observed to increase from an initial value of $Re_\tau \approx 190$, which was comparable to that of a rough bottom channel flow, to a final value of $Re_\tau \approx 300$. The increase of the fluid velocity fluctuations, as well as that of the bottom friction, as a function of time is in close correlation with the time evolution of the sediment bed height fluctuations, i.e. the increase of the amplitude of the patterns. It is well known that an increase of the bottom friction of a channel flow has a consequence of altering the mean velocity profile, when scaled in wall units (Jiménez, 2004; Pope, 2000). For the present simulated turbulent case, the degree of alteration is quantified by computing the roughness function ΔU^+ as a function of the roughness height of the patterns, i.e their amplitude normalized by viscous length scale $\langle A_h \rangle_D^+$. A linear relationship is recovered between ΔU^+ and $\langle A_h \rangle_D^+$.

In the laminar cases the volumetric particle flow rate (per unit span) is consistent with a cubic power law as a function of the Shields number; these values are found to be not far from those obtained in featureless bedload transport. The value corresponding to the turbulent case is sufficiently predicted by the transport law of Wong and Parker (2006) which is derived for turbulent flow in the presence of a plane mobile bed. It should be note that the evolution of 'vortex dunes' has indeed increased the net particle transport rate. But, at the same time, the bottom shear stress has also increased, with an overall negligible effect on the scaling law. The present results therefore seem to suggest that the presence of 'small dunes' as well as that of 'vortex dunes' up to the amplitudes encountered in the present simulations does not lead to a modification of the scaling of the net particle transport rate as a function of the (excess) shear stress at the bottom.

The turbulent flow field evolving over the 'vortex dunes' is statistically unsteady and two-dimensional. It features complex flow structures which exhibit strong correlation with the bottom topology of the sediment bed. Best (2005) has summarized the distinct flow structures as follows: an accelerating and a decelerating flow regions upstream and downstream of the dune crest respectively, a shear layer region downstream of the dune crest where flow separation occurs, a re-circulation region extending several dune heights downstream of the crest and bounded by the shear layer and a developing boundary layer region attached to the stoss side of the dune. In order to statistically characterize such a complex flow, we have introduced a 'dune-conditioned' spatial and temporal averaging of the flow field and particle motion, which takes into account the spatial and temporal variability of the sediment bed. The dune-conditioned averaging was then applied to the DNS data of the turbulent case in an observation interval of approximately 400 bulk time units near the end of the simulation. The mean fluid velocity and pressure fields exhibit strong correlation with the morphology of the dunes, featuring the above mentioned flow regions. DNS results are consistent with available experimental data (Charru and Franklin, 2012). Covariances of fluid velocity fluctuations are also investigated in detail. The turbulent kinetic energy as well as Reynolds shear stress are found to be maximum in the recirculation region downstream of the crest of the dunes. The values of the covariances attained reaching up to $6u_\tau^2$, $2u_\tau^2$, and $3u_\tau^2$ for the streamwise, wall-normal and spanwise components consecutively. The Reynolds shear stress attains values up to $2u_\tau^2$. These values are in good agreement with the data reported by Charru and Franklin (2012).

In addition to the fluid statistics, particle related quantities were also characterized. At, and above, the stoss side of the dunes, particle streamwise velocities were observed to be systematically smaller than that of the fluid. The observed velocity lag has been mainly attributed to the preferential location of particles with respect to the near wall high-/low-speed fluid streaks, similar observation as in rough-bottom or smooth channel flow (Chan-Braun, 2012; Kidanemariam et al., 2013). Covariances of particle velocity fluctuations exhibit similar spatial structure to that of fluid with a slight tendency of becoming more isotropic. It has been previously speculated that inter-particle collisions are responsible for the observed tendency (Kidanemariam et al., 2013). Additionally, the dune-conditioned averaging was utilized to study the spatial structure of the hydrodynamic and collision forces acting on the particles. Particles experience a positive hydrodynamic drag on the stoss side of the dunes upstream of the crests, while on the downstream side, a negative drag force is experienced. And the inter-particle collision forces generally counteracts the hydrodynamic forces. However, there are regions of the flow where a net acceleration and deceleration of particles is observed. These regions are strongly correlated with the dune geometry, with a net acceleration of particles upstream of the crest and a net deceleration in the downstream region.

It is well known that the mean total shear stress in particle laden flows is the sum of the fluid shear stresses as well as stress contributions arising from the fluid-particle interaction (see e.g. Righetti and Romano, 2004; Uhlmann, 2008; Kidanemariam et al., 2013). For flows over two-dimensional dunes, the fluid shear stress has three components; the viscous shear stress, the turbulent Reynolds shear stress and the dune shape induced stress contribution which arises as a result of the deflection of the streamlines of the mean flow (see e.g. Nikora et al., 2007b). In a statistically steady flow, the imposed pressure gradient is balanced by the sum of these three fluid shear stresses plus the stress resulting from the fluid-solid interaction. In the clear fluid region above the highest crest of the dunes, the fluid stress entirely balances the pressure gradient, while in the domain below the lowest trough of the dunes, the fluid-particle interaction stress entirely balances the pressure gradient. In between these two regions, all stress contributions are relevant. As a result of the contraction and expansion of the streamlines of the mean flow, the form-induced stress works against the turbulent Reynolds stress by transporting high-momentum fluid away from the bottom into the outer flow and vice-versa.

Of particular interest to the theoretical study of sedimentary patterns is the accurate determination of the spatial variation of the bottom boundary shear stress along the bedforms and its relation to the local volumetric particle flow rate. To this end, a method is presented which allows the determination of the local boundary shear stress from the integration of the mean momentum equation in a control volume moving at the propagation velocity of the dunes (cf. figure 6.19). The idea is similar to the von Kármán momentum integral approach commonly used to determine the boundary shear stress in the boundary layer flow community (Schlichting and Gersten, 2000; Pope, 2000). Results of such an analysis revealed that the boundary shear stress is maximum not at the crest of the dunes, but rather at a location upstream of the crests. In the stoss side of the dunes, the boundary shear stress attains values $1.5 - 1.8$ times the mean value, in good agreement with the available literature. On other hand, a negative value of shear stress is recovered downstream of the crest, consistent with the fact that the flow reverses in this region. Based on the cross-correlation between the dune geometry and boundary shear stress, the latter exhibits a positive phase shift of approximately $15D$. Such a phase shift is believed to arise from a combined effect of the form-induced pressure (which causes the accelerating/decelerating fluid regions) and fluid inertia (Charru et al., 2013; Andreotti and Claudin, 2013) and is claimed to be responsible for the instability of a sediment bed.

In the considered observation interval, in which the patterns migrate at an approximately constant propagation velocity and show negligible evolution, the volumetric particle flow rate evolves in phase with the sediment bed, with maximum value in the crest of the dunes and minimum value in the trough. This means that the boundary shear stress and the particle flow rate are out of phase by approximately $15D$. This shows that describing the local particle flow rate by a local shear stress is not adequate. The phase shift between the bottom shear stress and the particle flow rate is, in some studies, explained as the length scale needed for the particle flow rate to adopt to change in shear stress, termed saturation length scale L_{sat} (Charru et al., 2013). The outcome of the DNS data is valuable for evaluating models which take into account a relaxation effect by incorporate the saturation length scale.

Chapter 7

Conclusion

Under certain circumstances, a small perturbation of an initially flat subaqueous sediment bed will be amplified and give rise to wave-like patterns in the form of ripples or dunes which are commonly observed in rivers and coastal waters. Providing fundamental understanding of the mechanisms involved in the formation of these patterns is highly desirable. However, the complex interaction between the sediment particles and the driving turbulent shear flow as well as the dependence on multiple parameters has made this task very challenging to the present date. Our aim is to contribute to the understanding of the physical mechanisms involved in the processes of subaqueous bedload transport and ripple formation. To this end, we have numerically investigated the problem considering a statistically steady horizontal channel flow configuration, both in the laminar and turbulent regimes. The originality of our work lies on the adoption of high-fidelity simulations which resolve all the relevant scales of the flow problem, even the near field around the individual particles. Moreover, in order to realistically capture the formation of bedforms, we have adopted sufficiently large computational domains involving up to $O(10^9)$ grid nodes and up to $O(10^6)$ freely moving fully-resolved particles representing the bed Part of the present work has been praised by the community as "an excellent piece of work, whereby the complexity of the model has been pushed to a level never before reached in the field of sediment transport modelling" (Colombini, 2014).

Numerical approach

For the purpose of performing of the simulations, we have employed an existing fluid solver which features an immersed boundary technique for the accurate treatment of the moving fluid-solid interfaces (Uhlmann, 2005a). The algorithm was coupled with a collision model based upon the soft-sphere approach. The forces arising during solid-solid contact are expressed as a function of the overlap length, with an elastic and a damping normal force component, as well as a damping tangential force, limited by a Coulomb friction law. Since the characteristic collision time is typically orders of magnitude smaller than the time step of the flow solver, the numerical integration of the equations for the particle motion is carried out adopting a sub-stepping technique, freezing the hydrodynamic forces acting upon the particles between successive flow field updates.

The collision strategy was validated with respect to the test case of gravity-driven motion of a single sphere colliding with a horizontally oriented plane wall in a viscous fluid. Simulations over a range of Stokes numbers have been performed. It was found that the present collision strategy works correctly when coupled to the particulate flow solver, yielding values for the effective coefficient of restitution in close agreement with reference data from experimental measurements.

Erosion of a sediment bed sheared by laminar channel flow

In the intermediate phase of our work, we have addressed the problem of bedload transport in a laminar channel flow. The aim of this work was, first, to demonstrate the adequacy of our numerical approach to simulate the formation of patterns and, second, to provide new data for the purpose of studying the bedload problem by itself. To this end, we have performed direct numerical simulations of the erosion of a thick bed of mobile sediment particles by a laminar channel flow, similar to the experiment of Aussillous et al. (2013), for 24 different parameter values covering a broad range of the relevant control parameters of the problem. The DNS-DEM method was found to provide results which are fully consistent with the available data from the reference experiment with respect to the variation of the particle flow rate, thickness of the mobile layer as well as fluid and particle velocities as a function of Shields number (or the non-dimensional flow rate). Based upon the DNS data, it was shown that the particle flow rate varies with the third power of the Shields number, once the threshold value of the Shields number is exceeded, consistent with experimental observations.

Moreover, it was shown that the fluid mean velocity exhibits a parabolic profile in most part of the clear fluid region. The fluid and particle velocities inside the mobile granular layer were fully described by defining appropriate velocity and length scales, namely the value of the mean fluid velocity at the fluid-bed interface and a length scale termed 'characteristic half thickness', which corresponds to the wall-normal offset from the fluid-bed interface to the location where the mean fluid velocity reduces to 50% of the chosen velocity scale. It turns out that the fluid mean velocity at the interface exhibits a quadratic relationship with the Shields number (or the non-dimensional fluid flow rate) while the characteristic half thickness exhibits a linear variation. An empirical exponential curve, fully defined by the mentioned length and velocity scales, is proposed which satisfactorily fits the normalized fluid velocity profiles inside the granular layer. Finally, the contribution of the hydrodynamic and inter-particle collision forces to the momentum transfer between the two phases was addressed. The effective coefficient of friction as well as the solid volume fraction of a submerged sediment bed were observed to be well described by the dimensionless viscous number proposed by Boyer et al. (2011).

Pattern formation over a sheared subaqueous sediment bed in channel flow

Following the successful simulation of the case of featureless bedload transport, we have performed several direct numerical simulations of the formation of bedforms over an erodible bed, covering four sets of parameter values both in the laminar and turbulent regimes. Consistent with the regime classification of Ouriemi et al. (2009a), one of the laminar cases evolved without the formation of patterns while the other two laminar cases lead to the formation of 'small dunes'. The simulation performed under turbulent flow conditions exhibited 'vortex dunes'. The spatial and temporal variation of the sediment bed height, which was extracted from the spanwise-averaged solid volume fraction, was analyzed in order to investigate the time evolution of characteristics of the emerged patterns. The amplitude of the 'vortex dunes' was observed to grow at an initial rate with is approximately ten times larger than that of the 'small dunes' consistent with available experimental data. The observed initial wavelengths of the patterns, which were in the range 100–150D for all the cases, as well as their propagation velocity were also in good agreement with previous experimental observations. The turbulent case simulation is repeated several times by keeping all relevant parameter values identical but minimizing the computational domain in the streamwise direction in order to find the minimal box length which will accommodate the lower bound of the unstable pattern wavelength at the instants of the initial bed instability, a strategy similar to the minimal flow unit of Jiménez and Moin (1991). The outcome of the different simulations indicated that the cutoff length for pattern formation lies in the range 78–102D, or in terms of the half channel height, 3–4H, for the chosen parameter values.

The influence of the computational domain size on the selection and evolution of the most amplified wavelength was further studied by performing one simulation adopting a superior domain size. The box was able to accommodate eleven initial wavelengths and allowed the determination of the most amplified initial wavelength to be determined with sufficient accuracy. It turns out that the most amplified mode has a wavelength $\lambda_c \approx 112D \approx 4.36H$. It can be concluded that a computational domain length with $L_x/\lambda_c = O(1)$, due to the discreteness of the resolved harmonics, severely constrains the natural selection mechanism of the initial pattern wavelength.

Fluid flow and particle motion over subaqueous patterns in channel flow

The flow field, as well as the associated particle motion, over the subaqueous sediment patterns was rigorously studied. Influence of the evolving sediment bed on the bulk characteristics of the flow was addressed by scrutinizing the evolution of the fluid velocity fluctuations and mean bottom shear stress. It is found that the evolution of the 'small dunes' has almost no influence on the bulk statistics of the driving laminar flow, while the 'vortex dunes' are observed to substantially alters the statistics of the turbulent flow. In the laminar cases the volumetric particle flow rate (per unit span) is consistent with a cubic power law as a function of the Shields number; these values are found to be not far from those obtained in featureless bedload transport. The value corresponding to the turbulent case is sufficiently predicted by the transport law of Wong and Parker (2006) which is derived for turbulent flow in the presence of a plane mobile bed. The present results therefore seem to suggest that the presence of 'small dunes' as well as that of 'vortex dunes' up to the amplitudes encountered in the present simulations does not lead to a modification of the scaling of the net particle transport rate as a function of the (excess) shear stress at the bottom.

The highly complex turbulent flow evolving over the 'vortex dunes' exhibit strong correlation with the bottom topology of the sediment bed. For the purpose of characterizing the flow, dune-conditioned spatial and temporal averaging of the flow field and particle motion, which takes into account the spatial and temporal variability of the sediment bed was performed. We were able to educe the properties of the statistically two-dimensional mean flow by such an analysis. Consistent with what is reported in the literature, the turbulent kinetic energy as well as Reynolds shear stress exhibited maximum values in the recirculation region downstream of the crest of the dunes. Moreover, detailed dune-conditioned statistics of particle velocities as well as forces acting upon particles were performed. The accurate determination of the spatial variation of the bottom boundary shear stress along the bedforms and its relation to the local volumetric particle flow rate is very crucial in the study of sedimentary patterns. To this end, a method is presented which allows the determination of the local boundary shear stress from the integration of the mean momentum equation of the time- and space-varying flow. Results of such an analysis revealed that the boundary shear stress is maximum at a location upstream of the crests exhibiting a positive phase shift with respect to the dune geometry. Such a phase shift is claimed to be responsible for the instability of a sediment bed (Charru et al., 2013). As a consequence of the sediment continuity, the volumetric particle flow rate evolves in phase with the boundary and thus lags the shear stress by the same phase shift amount. This shows that relating the local particle flow rate and a local shear stress by an algebraic expression is not adequate.

Appendix A

Linear mass spring damper system

We consider a simple mass spring damper system consisting an object of mass M, approaching a flat plate which is attached to a stationary wall by a linear spring, with a stiffness parameter k_n, and a linear damper, with a damping coefficient c_{dn} (cf. figure A.1). We choose a coordinate system such that the equilibrium position of the plate before impact is located at $x = 0$. We assume no external force and a constant approach velocity u_0 of the object. Upon impact, the flat plate will be displaced from its equilibrium position and the system can be described by the following ODE.

$$\ddot{x} - \frac{c_{dn}}{M}\dot{x} + \frac{k_n}{M}x = 0 \,. \tag{A.1}$$

Equation (A.1) simply tells us that the linear spring exerts a repulsive force which is proportional but opposite in direction to the displacement and the linear damper exerts a force which is proportional to but opposite in direction to the rate of change of the displacement. Let us define two parametes, namely the undamped natural frequency of the system ω_0 and the damping ratio ζ given as

$$\omega_0 = \sqrt{k_n/M} \quad \text{and} \quad \zeta = \frac{c_{dn}}{2\sqrt{k_n M}} \tag{A.2}$$

After substitution, equation A.1 becomes

$$\ddot{x} - 2\zeta\omega_0\dot{x} + \omega_0^2 x = 0 \,. \tag{A.3}$$

The above system, under the condition that the object with mass M remains 'glued' to the plate for all t after impact, has the following behavior which depend on the value of ζ. If $\zeta = 1$, the system is said to be 'critically damped' and returns to equilibrium as quickly as possible without oscillating. If $\zeta > 1$, the system is said to be 'overdamped' and returns to equilibrium without oscillating and the higher the value of ζ, the slower it will return to equilibrium. If $0 < \zeta < 1$, the system is said to be 'underdamped' and oscillates with a damped frequency and a time-decreasing amplitude. If $\zeta = 0$, the system is said to be 'undamped' and the system oscillates at its natural frequency ω_0 forever.

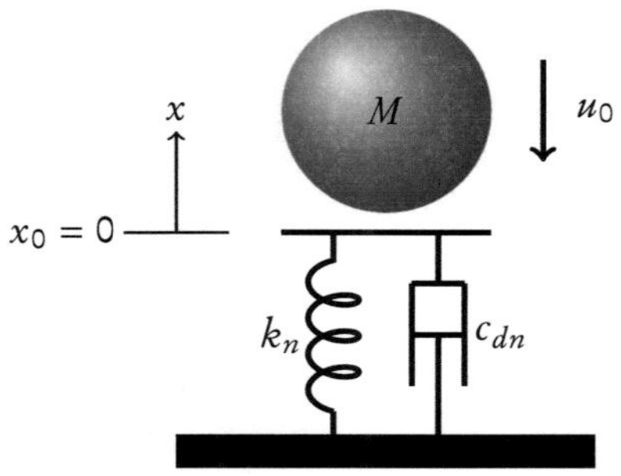

Figure A.1: Schematic diagram of a mass spring damper system

For our interest, i.e. for $0 < \zeta < 1$, the system (A.3) has the following solution:

$$x(t) = e^{-\zeta\omega_0 t}(A\cos(\omega_d t) + B\sin(\omega_d t)) \tag{A.4a}$$

$$\dot{x}(t) = e^{-\zeta\omega_0 t}\Big[\omega_d\big(B\cos(\omega_d t) - A\sin(\omega_d t)\big) - \zeta\omega_0\big(A\cos(\omega_d t) + B\sin(\omega_d t)\big)\Big] \tag{A.4b}$$

where

$$\omega_d = \omega_0\sqrt{1-\zeta^2}; \quad A = x(0) \quad \text{and} \quad B = \frac{1}{\omega_d}(A\zeta\omega_0 + \dot{x}(0)). \tag{A.5}$$

Note that, ω_d is the damped frequency of the system. After substituting the initial conditions, equation A.4 can be further reduced to

$$x(t) = \frac{u_0}{\omega_d}e^{-\zeta\omega_0 t}\sin(\omega_d t) \tag{A.6a}$$

$$\dot{x}(t) = u_0 e^{-\zeta\omega_0 t}\Big[\cos(\omega_d t) - \frac{\zeta\omega_0}{\omega_d}\sin(\omega_d t)\Big] \tag{A.6b}$$

Equation (A.6) describes the oscillation of the system only if it is guarantied that the object remains attached to the plate for all t after the initial impact. But when the impacting object is not attached to the plate, equation (A.6) is valid only in the time interval between the initial time $t = 0$ and the time at which $x(t)$ is zero again (say T_c). After this time, the spring and damper will no longer constrain the motion of the object and it will bounce back with a constant velocity $u_1 = u(T_c)$ (cf. figure A.2). It can be shown that $u_1 = -u_0 e^{-\pi\zeta\omega_0/\omega_d}$. Thus, the restitution coefficient ε, defined as the absolute value of the ratio of the post-collision to pre-collision velocities u_1 and u_0 respectively, is given by

$$\varepsilon = e^{-\pi\zeta\omega_0/\omega_d}\,. \tag{A.7}$$

From equation (A.7) it can be easily shown that, for $0 < \zeta < 1$,

$$\zeta = -\frac{\ln\varepsilon}{\sqrt{\pi^2 + \ln^2\varepsilon}}\,. \tag{A.8}$$

This means, for the system to rebound the object with a given restitution coefficient ε, the damper coefficient c_{nd} has to be

$$c_{dn} = -2\sqrt{Mk_n}\frac{\ln\varepsilon}{\sqrt{\pi^2 + \ln^2\varepsilon}}\,. \tag{A.9}$$

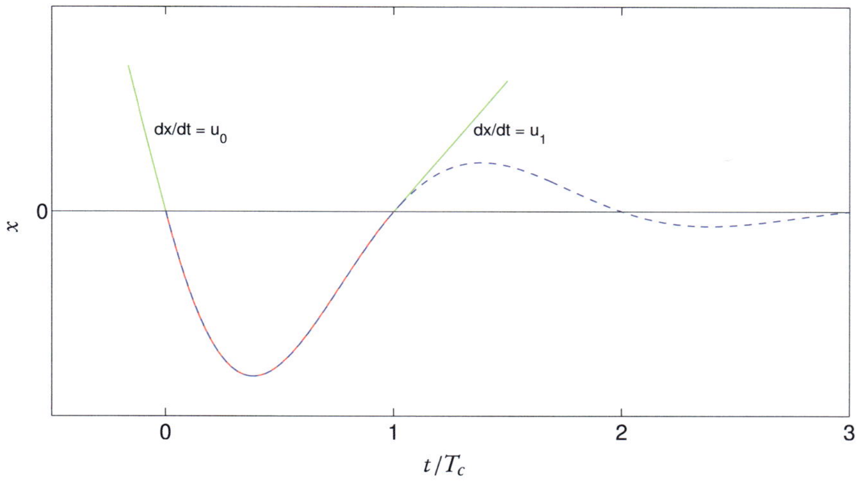

Figure A.2: Time evolution of the displacement of an attached mass spring damper system (blue-dashed curve); the trajectory of an object (corresponding to its contacting edge) hitting a spring damper system with impact velocity u_0 at time t_0 and rebounding with a velocity u_1 at time T_c (green and red curves). The red curve alone represents the trajectory during which the mass stays in contact with the spring damper system.

The duration $T_c = \pi/\omega_d$, which is half the oscillation period of the system is a crucial parameter in determining the time step interval which should be adopted when numerically solving the system ODE (A.1). It can be rearranged into

$$T_c = \frac{2\pi M}{\sqrt{4Mk_n - c_{dn}^2}} \,. \tag{A.10}$$

In the context of the linear mass spring damper model, an object of mass M, hitting a stationary flat wall, will stay in contact with wall for a duration of T_c before it gets rebounded with an opposite and damped velocity. Note that T_c does not depend on the magnitude of the impact velocity.

Appendix B

Sub-stepping

In the context of the soft-sphere approach, the DEM solver requires very small time steps to resolve the contact forces during particle-particle or particle-wall collisions (due to the very small thickness of the admitted overlap), and when coupled with the DNS code it typically imposes a rigid constraint on the computational time step. In order to avoid this restriction, one approach is to apply the sub-stepping strategy by which the Newton equations are explicitly solved at a smaller time step than that used for solving the Navier-Stokes equations (see e.g. Wachs, 2009).

Runge-Kutta levels

In order to apply the sub-stepping strategy, it becomes necessary to determine the time levels Δt^k of each of the three steps of the Runge-Kutta time integration scheme within one time advancement from level n to level $n+1$. Previously, in the framework of the direct forcing method, Uhlmann (2005a) has discretized the Newton equations for particle motion by applying the following three-step Runge-Kutta method, consistent with the flow solver therein:

$$\mathcal{F}^{*(k)} = f^*(\boldsymbol{x}_p^{*(k-1)}, \boldsymbol{u}_p^{*(k-1)}, \boldsymbol{u}_f^{*(k)}) \tag{B.1}$$

$$\boldsymbol{a}_p^{*(k)} = -\frac{\rho_f}{V_c(\rho_p - \rho_f)} \mathcal{F}^{*(k)} \tag{B.2}$$

$$\boldsymbol{u}_p^{*(k)} = \boldsymbol{u}_p^{*(k-1)} + \Delta t\, \boldsymbol{a}_p^{*(k)} \tag{B.3}$$

$$\boldsymbol{x}_p^{*(k)} = \boldsymbol{x}_p^{*(k-1)} + \Delta t \alpha_k (\boldsymbol{u}_p^{*(k)} + \boldsymbol{u}_p^{*(k-1)}) \tag{B.4}$$

where the superscript * indicates a numerical value and k is the index of the Runge-Kutta step. $\boldsymbol{a}_p^{*(k)}$, $\boldsymbol{u}_p^{*(k)}$ and $\boldsymbol{x}_p^{*(k)}$ are the particle's translational acceleration due to hydrodynamic force (excluding gravity forces for the moment), translational velocity and center position vector respectively. Values at the step $k = 0$ and $k = 3$ correspond to those at time t^n and $t^{n+1} = t^n + \Delta t$ respectively. The full Runge-kutta set of coefficients adopted in the DNS code are the following:

$$\alpha_k = \left\{\frac{4}{15}, \frac{1}{15}, \frac{1}{6}\right\}, \quad \gamma_k = \left\{\frac{8}{15}, \frac{5}{12}, \frac{3}{4}\right\}, \quad \zeta_k = \left\{0, -\frac{17}{60}, -\frac{5}{12}\right\}. \tag{B.5}$$

Let the exact position and velocity at the k-th Runge-Kutta time level Δt_k be denoted as $\boldsymbol{x}_p^{(k)}$ and $\boldsymbol{u}_p^{(k)}$ respectively. The difference between the exact and the numerically computed values (the error) is given by

$$\epsilon_x^{(k)} = \boldsymbol{x}_p^{(k)} - \boldsymbol{x}_p^{*(k)} \tag{B.6}$$

$$\epsilon_u^{(k)} = \boldsymbol{u}_p^{(k)} - \boldsymbol{u}_p^{*(k)}. \tag{B.7}$$

In order to determine Δt_k (which is our aim), one would obtain the time levels which yield $\epsilon_x^{(k)} = \epsilon_u^{(k)} = 0$ to within the achievable order of accuracy. Defining $\chi_k = \Delta t_k / \Delta t$, the Taylor expansions of $\boldsymbol{u}_p^{(n)}$ and $\boldsymbol{x}_p^{(n)}$ around the time Δt_k are given by:

$$\boldsymbol{u}_p^{(n)} = \boldsymbol{u}_p^{(k)} - \chi_k \Delta t \boldsymbol{u}_p'^{(k)} + \frac{1}{2}\chi_k^2 \Delta t^2 \boldsymbol{u}_p''^{(k)} + O(\Delta t^3) \tag{B.8}$$

$$\boldsymbol{x}_p^{(n)} = \boldsymbol{x}_p^{(k)} - \chi_k \Delta t \boldsymbol{u}_p^{(k)} + \frac{1}{2}\chi_k^2 \Delta t^2 \boldsymbol{u}_p'^{(k)} + O(\Delta t^3) \tag{B.9}$$

Successively substituting of the Runge-kutta levels:

1. first level ($k = 1$):

$$\boldsymbol{a}_p^{*(1)} = f^*(\boldsymbol{x}_p^{(n)}, \boldsymbol{u}_p^{(n)}, \boldsymbol{u}_f^{*(1)}, \rho_p, \rho_f, V_p) \tag{B.10}$$

$$\boldsymbol{u}_p^{*(1)} = \boldsymbol{u}_p^{(n)} + \boldsymbol{a}_p^{*(1)} \Delta t \tag{B.11}$$

$$\boldsymbol{x}_p^{*(1)} = \boldsymbol{x}_p^{(n)} + 2\alpha_1 \boldsymbol{u}_p^{*(1)} \Delta t - \alpha_1 \boldsymbol{a}_p^{*(1)} \Delta t^2 \tag{B.12}$$

From equations (B.11) and (B.8):

$$\begin{aligned} \epsilon_u^{(1)} &= \boldsymbol{u}_p^{(1)} - \boldsymbol{u}_p^{*(1)} \\ &= \underbrace{\left(\chi_1 \boldsymbol{u}_p'^{(1)} - \boldsymbol{a}_p^{*(1)}\right)}_{=\Pi_1} \Delta t + O(\Delta t^2) \\ &= O(\Delta t^2) \quad \text{if } \Pi_1 = 0 \quad \Rightarrow \boldsymbol{u}_p'^{(1)} = \frac{1}{\chi_1} \boldsymbol{a}_p^{*(1)} \end{aligned} \tag{B.13}$$

From equations (B.12) and (B.9):

$$\begin{aligned} \epsilon_x^{(1)} &= \boldsymbol{x}_p^{(1)} - \boldsymbol{x}_p^{*(1)} \\ &= \underbrace{\chi_1 \epsilon_u^{(1)} \Delta t}_{O(\Delta t^3)} + \underbrace{(\chi_1 - 2\alpha_1) \boldsymbol{u}_p^{*(1)}}_{\Pi_2} \Delta t + \frac{1}{2} \underbrace{(2\alpha_1 - \chi_1)}_{\Pi_3} \boldsymbol{a}_p^{*(1)} \Delta t \check{s} + O(\Delta t^3) \\ &= O(\Delta t^3) \text{ if } \Pi_2 = 0 \text{ and } \Pi_3 = 0 \quad \Rightarrow \chi_1 = 2\alpha_1 \end{aligned} \tag{B.14}$$

Thus, the values of the particle velocity and position vectors computed at the first Runge-Kutta step are approximations to the corresponding exact values at time

$$t^{(1)} = t^{(n)} + 2\alpha_1 \Delta t \; . \tag{B.15}$$

The achieved order of accuracies (of the time derivatives) are $O(\Delta t)$ for the velocity and $O(\Delta t^2)$ for the position. The hydrodynamic force responsible for advancing the particle from time $t^{(n)}$ to $t^{(1)}$ becomes

$$\frac{\boldsymbol{u}_p^{(1)} - \boldsymbol{u}_p^{(n)}}{\chi_1 \Delta t} = \frac{1}{\chi_1} \boldsymbol{a}_p^{*(1)} + O(\Delta t) \tag{B.16}$$

2. second level ($k = 2$):

$$\boldsymbol{a}_p^{*(2)} = f^*(\boldsymbol{x}_p^{*(1)}, \boldsymbol{u}_p^{*(1)}, \boldsymbol{u}_f^{*(2)}, \rho_p, \rho_f, V_p) \tag{B.17}$$

$$\boldsymbol{u}_p^{*(2)} = \boldsymbol{u}_p^{(n)} + \left(\boldsymbol{a}_p^{*(1)} + \boldsymbol{a}_p^{*(2)}\right)\Delta t \tag{B.18}$$

$$\boldsymbol{x}_p^{(2)} = \boldsymbol{x}_p^{(n)} + 2(\alpha_1 + \alpha_2)\boldsymbol{u}_p^{*(2)}\Delta t - \left(\alpha_1 \boldsymbol{a}_p^{*(1)} + (2\alpha_1 + \alpha_2)\boldsymbol{a}_p^{*(2)}\right)\Delta t^2 \tag{B.19}$$

From equations (B.18) and (B.8):

$$\begin{aligned}
\epsilon_u^{(2)} &= \boldsymbol{u}_p^{(2)} - \boldsymbol{u}_p^{*(2)} \\
&= \underbrace{\left(\chi_2 \boldsymbol{u}_p'^{(2)} - \boldsymbol{a}_p^{*(1)} - \boldsymbol{a}_p^{*(2)}\right)}_{\Pi_4} \Delta t + O(\Delta t^2) \\
&= O(\Delta t^2) \text{ if } \Pi_4 = 0 \quad \Rightarrow \quad \boldsymbol{u}_p'^{(2)} = \frac{1}{\chi_2}\left(\boldsymbol{a}_p^{*(1)} + \boldsymbol{a}_p^{*(2)}\right)
\end{aligned} \tag{B.20}$$

From equation (B.19) and (B.9):

$$\begin{aligned}
\epsilon_x^{(2)} &= \boldsymbol{x}_p^{(2)} - \boldsymbol{x}_p^{*(2)} \\
&= \underbrace{\chi_2 \epsilon_u^{(2)} \Delta t}_{O(\Delta t^3)} + \underbrace{(\chi_2 - 2(\alpha_1 + \alpha_2))}_{\Pi_5} \boldsymbol{u}_p^{*(2)} \Delta t \\
&\quad + \underbrace{\left\{ \left(\frac{1}{2}\chi_2 + \alpha_1\right)\boldsymbol{a}_p^{*(1)} + \left(\frac{1}{2}\chi_2 + 2\alpha_1 + \alpha_2\right)\boldsymbol{a}_p^{*(2)} + \right\}}_{\Pi=6} \Delta t^2 + O(\Delta t^3) \\
&= O(\Delta t^2) \quad \text{if } \Pi_5 = 0 \quad \Rightarrow \chi_2 = 2(\alpha_1 + \alpha_2)
\end{aligned} \tag{B.21}$$

Thus, the values of the particle velocity and position vectors computed at the second Runge-Kutta step are approximations to the corresponding exact values at time

$$t^{(2)} = t^{(n)} + (2\alpha_1 + 2\alpha_2)\Delta t \ . \tag{B.22}$$

The achieved order of accuracies (of the time derivatives) are $O(\Delta t)$ for both the velocity and the position. The hydrodynamic force responsible for advancing the particle from time $t^{(n)}$ to $t^{(2)}$ is given by (after Taylor expansion)

$$\frac{\boldsymbol{u}_p^{(2)} - \boldsymbol{u}_p^{(n)}}{\chi_2 \Delta t} = \frac{1}{\chi_2}\left(\boldsymbol{a}_p^{*(1)} + \boldsymbol{a}_p^{*(2)}\right) + O(\Delta t) \tag{B.23}$$

For the purpose of the sub-stepping, a more relevant force is the hydrodynamic force responsible for advancing the particle from time $t^{(1)}$ to $t^{(2)}$ which is given by:

$$\frac{\boldsymbol{u}_p^{(2)} - \boldsymbol{u}_p^{(1)}}{(\chi_2 - \chi_1)\Delta t} = \frac{\chi_2 \boldsymbol{u}_p'^{(2)} - \chi_1 \boldsymbol{u}_p'^{(1)}}{\chi_2 - \chi_1} + O(\Delta t) = \frac{1}{\chi_2 - \chi_1}\boldsymbol{a}_p^{*(2)} + O(\Delta t) \tag{B.24}$$

3. third level ($k = 3$):

$$\boldsymbol{a}_p^{*(3)} = f^*(\boldsymbol{x}_p^{*(2)}, \boldsymbol{u}_p^{*(2)}, \boldsymbol{u}_f^{*(3)}, \rho_p, \rho_f, V_p) \tag{B.25}$$

$$\boldsymbol{u}_p^{*(3)} = \boldsymbol{u}_p^{(n)} + \left(\boldsymbol{a}_p^{*(1)} + \boldsymbol{a}_p^{*(2)} + \boldsymbol{a}_p^{*(3)}\right)\Delta t \tag{B.26}$$

$$\begin{aligned} \boldsymbol{x}_p^{*(3)} &= \boldsymbol{x}_p^{(n)} + 2(\alpha_1 + \alpha_2 + \alpha_3)\boldsymbol{u}_p^{*(3)}\Delta t \\ &\quad -\alpha_1 \boldsymbol{a}_p^{*(1)}\Delta t^2 + (-2\alpha_1 + \alpha_2)\boldsymbol{a}_p^{*(2)}\Delta t^2 - (2\alpha_1 + 2\alpha_2 + \alpha_3)\boldsymbol{a}_p^{*(3)}\Delta t^2 \end{aligned} \tag{B.27}$$

From equations (B.26) and (B.8):

$$\begin{aligned} \epsilon_u^{(3)} &= \boldsymbol{u}_p^{(n+1)} - \boldsymbol{u}_p^{*(3)} \\ &= \underbrace{\left(\boldsymbol{u}_p'^{(n+1)} - \boldsymbol{a}_p^{*(1)} - \boldsymbol{a}_p^{*(2)} - \boldsymbol{a}_p^{*(3)}\right)}_{\Pi_7}\Delta t + O(\Delta t^2) \\ &= O(\Delta t^2) \quad \text{if} \quad \Pi_7 = 0 \quad \Rightarrow \quad \boldsymbol{u}_p'^{(n+1)} = \boldsymbol{a}_p^{*(1)} + \boldsymbol{a}_p^{*(2)} + \boldsymbol{a}_p^{*(3)} \end{aligned} \tag{B.28}$$

From equations (B.27) and (B.9):

$$\begin{aligned} \epsilon_x^{(3)} &= \boldsymbol{x}_p^{(n+1)} - \boldsymbol{x}_p^{*(3)} \\ &= \epsilon_u^{(2)}\Delta t + (1 - 2(\alpha_1 + \alpha_2 + \alpha_3))\boldsymbol{u}_p^{*(3)}\Delta t \\ &\quad + \Big\{\left(\alpha_1 - \tfrac{1}{2}\right)\boldsymbol{a}_p^{*(1)} + \left(2\alpha_1 - \alpha_2 - \tfrac{1}{2}\right)\boldsymbol{a}_p^{*(2)} + \left(2\alpha_1 + 2\alpha_2 + \alpha_3 - \tfrac{1}{2}\right)\boldsymbol{a}_p^{*(3)}\Big\}\Delta t^2 + O(\Delta t^3) \\ &= O(\Delta t^2) \end{aligned} \tag{B.29}$$

Thus, at the end of one time step, the achieved order of accuracies (of the time derivatives) are $O(\Delta t)$ for both the velocity and the position with the total hydrodynamic force to advance from time level $t^{(n)}$ to level $t^{(n+1)}$ given by

$$\frac{\boldsymbol{u}_p^{(n+1)} - \boldsymbol{u}_p^{(n)}}{\Delta t} = \boldsymbol{a}_p^{*(1)} + \boldsymbol{a}_p^{*(2)} + \boldsymbol{a}_p^{*(3)} + O(\Delta t) \tag{B.30}$$

and the force responsible to advance the particle from time level $t^{(2)}$ to level $t^{(n+1)}$

$$\frac{\boldsymbol{u}_p^{(n+1)} - \boldsymbol{u}_p^{(2)}}{(1 - \chi_2)\Delta t} = \frac{1}{1 - \chi_2}\boldsymbol{a}_p^{*(3)} + O(\Delta t) \tag{B.31}$$

Sub-stepping

Suppose we want to discretize the Newton equations at a sufficiently small time step Δt_{sub} such that $\Delta t = N_{sub}\Delta t_{sub}$, where Δt is the time step adopted to integrate the Navier-Stokes equations and N_{sub} is an integer number of sub-steps. Thus the fraction of Δt and the number of collision sub-steps per one Runge-kutta sub-step k is given by

$$\xi_k = \frac{\Delta t^k - \Delta t^{k-1}}{\Delta t} \tag{B.32}$$

$$N_{sub}^k = \xi_k N_{sub} \tag{B.33}$$

Note that N_{sub} has to be multiple of 15 in order to insure that N_{sub}^k is an integer. Any time integration scheme could be adopted for the advancement of the particle's translational velocities and positions within a collision sub-step. Here, we adopt the same Runge-kutta time integration scheme as was used in the DNS code by Uhlmann (2005a). Thus,

$$\frac{\boldsymbol{u}_p^{*(k^s)} - \boldsymbol{u}_p^{*(k^s-1)}}{\Delta t_{sub}} = \frac{-2\alpha_{k^s}}{\xi_k} \frac{\rho_f \mathcal{F}^{(k)}}{(\rho_p - \rho_f)V_p} + \frac{\gamma_{k^s}\mathbf{F}_c^{(k^s)} + \zeta_{k^s}\mathbf{F}_c^{(k^s-1)}}{(\rho_p - \rho_f)V_p} + 2\alpha_{k^s}\mathbf{g} \tag{B.34}$$

$$\frac{\boldsymbol{x}_p^{(k^s)} - \boldsymbol{x}_p^{(k^s-1)}}{\Delta t_{sub}} = \alpha_{k^s}(\boldsymbol{u}_p^{(k^s)} + \boldsymbol{u}_p^{(k^s-1)}) \tag{B.35}$$

where k^s is a Runge-Kutta step within one collision sub-step. Note that, $\mathcal{F}^{(k)}/\xi_k$ is constant during all the sub-steps N_{sub}^k within one "outer" Runge-Kutta step k. Thus it has a coefficient $2\alpha_{k^s}$ similar as for the gravity force.

The angular positions and angular velocities of the spherical particle are integrated in a similar fashion.

Appendix C

Momentum balance

Reynolds-averaged equations in a moving inertial frame

The momentum equation 2.2b can be written as (assuming $\rho_f = 1$):

$$\frac{\partial \boldsymbol{u}}{\partial t} + (\boldsymbol{u} \cdot \nabla)\boldsymbol{u} = -\nabla p + \nabla \overline{\overline{\boldsymbol{\tau}}} + \boldsymbol{f} \ . \tag{C.1}$$

where $\overline{\overline{\boldsymbol{\tau}}}$ is the viscous shear stress tensor. Let us consider an inertial frame of reference $\widehat{E}$ moving at a constant velocity $\boldsymbol{u}_D$ relative to the stationary frame. The transformations between the coordinate system are denoted as follows (see e.g. Pope, 2000):

$$\widehat{\boldsymbol{x}} = \boldsymbol{x} - \boldsymbol{u}_D t \ ; \quad \widehat{t} = t \ ; \quad \widehat{\boldsymbol{u}}(\widehat{\boldsymbol{x}},\widehat{t}) = \boldsymbol{u}(\boldsymbol{x},t) - \boldsymbol{u}_D \ . \tag{C.2}$$

Moreover, since the transformation is between two inertial frames, we have

$$\widehat{p} = p \ ; \quad \widehat{f} = f \ . \tag{C.3}$$

From chain rule, it follows that

$$\frac{\partial u_i}{\partial t} = \frac{\partial u_i}{\partial \widehat{t}}\frac{\partial \widehat{t}}{\partial t} + \frac{\partial u_i}{\partial \widehat{x}_j}\frac{\partial \widehat{x}_j}{\partial t} \quad \text{and} \quad \frac{\partial u_i}{\partial x_j} = \frac{\partial u_i}{\partial \widehat{t}}\frac{\partial \widehat{t}}{\partial x_j} + \frac{\partial u_i}{\partial \widehat{x}_k}\frac{\partial \widehat{x}_k}{\partial x_j} \tag{C.4}$$

where summation is implied over repeated indices. From (C.4) it can be shown that (see e.g. Pope, 2000):

$$\frac{\partial \boldsymbol{u}}{\partial t} = \frac{\partial \widehat{\boldsymbol{u}}}{\partial \widehat{t}} - (\boldsymbol{u}_D \cdot \widehat{\nabla})\widehat{\boldsymbol{u}} \quad \text{and} \quad \nabla \boldsymbol{u} = \widehat{\nabla}\widehat{\boldsymbol{u}} \tag{C.5}$$

Thus, transformation of (C.1) into $\widehat{E}$ yields:

$$\frac{\partial \widehat{\boldsymbol{u}}}{\partial \widehat{t}} + (\widehat{\boldsymbol{u}} \cdot \widehat{\nabla})\widehat{\boldsymbol{u}} = -\widehat{\nabla}\widehat{p} + \widehat{\nabla} \cdot \widehat{\overline{\overline{\boldsymbol{\tau}}}} + \widehat{\boldsymbol{F}} \ . \tag{C.6}$$

Now, let us consider a control volume $\mathcal{V}$ of constant shape, enclosed by a surface $\mathcal{S}$, attached to the moving frame of reference. That is, a control volume which is moving at a velocity of $\boldsymbol{u}_D$ with respect to E but is stationary with respect to $\widehat{E}$. Integrating (C.6) over $\mathcal{V}$ and applying Gauss' theorem in order to transform the volume integral into surface integral, we get:

$$\frac{\partial}{\partial \widehat{t}} \int_{\mathcal{V}} \widehat{\boldsymbol{u}} \mathrm{d}V + \oint_{\mathcal{S}} (\widehat{\boldsymbol{u}} \otimes \widehat{\boldsymbol{u}}) \cdot \widehat{\boldsymbol{n}} \mathrm{d}S = - \oint_{\mathcal{S}} \widehat{p} \overline{\overline{\boldsymbol{I}}} \cdot \widehat{\boldsymbol{n}} \mathrm{d}S + \oint_{\mathcal{S}} \widehat{\overline{\overline{\tau}}} \cdot \widehat{\boldsymbol{n}} \mathrm{d}S + \int_{\mathcal{V}} \widehat{f} \mathrm{d}V \,. \tag{C.7}$$

where $\widehat{\boldsymbol{n}}$ is a unit normal vector at the surface $\mathcal{S}$ and pointing outward, $\overline{\overline{\boldsymbol{I}}}$ is the identity tensor, and $\widehat{\boldsymbol{u}} \otimes \widehat{\boldsymbol{u}} \equiv \widehat{u}_i \widehat{u}_j$ is a diadic tensor.

Performing standard Reynolds averaging by decomposing the flow variables into a mean and a fluctuating components

$$\widehat{\boldsymbol{u}} = \langle \widehat{\boldsymbol{u}} \rangle + \widehat{\boldsymbol{u}}' \,, \quad \widehat{p} = \langle \widehat{p} \rangle + \widehat{p}' \,, \quad \widehat{f} = \langle \widehat{f} \rangle + \widehat{f}' \,, \tag{C.8}$$

the Reynolds averaged integral momentum equation becomes:

$$\frac{\partial}{\partial \widehat{t}} \int_{\mathcal{V}} \langle \widehat{\boldsymbol{u}} \rangle \mathrm{d}V + \oint_{\mathcal{S}} (\langle \widehat{\boldsymbol{u}} \rangle \otimes \langle \widehat{\boldsymbol{u}} \rangle) \cdot \widehat{\boldsymbol{n}} \mathrm{d}S + \oint_{\mathcal{S}} (\langle \widehat{\boldsymbol{u}}' \otimes \widehat{\boldsymbol{u}}' \rangle) \cdot \widehat{\boldsymbol{n}} \mathrm{d}S$$
$$= - \oint_{\mathcal{S}} \langle \widehat{p} \rangle \overline{\overline{\boldsymbol{I}}} \cdot \widehat{\boldsymbol{n}} \mathrm{d}S + \oint_{\mathcal{S}} \langle \widehat{\overline{\overline{\tau}}} \rangle \cdot \widehat{\boldsymbol{n}} \mathrm{d}S + \int_{\mathcal{V}} \langle \widehat{\boldsymbol{F}} \rangle \mathrm{d}V \tag{C.9}$$

Momentum balance in a moving inertial control volume

In this section, we will evaluate equation (C.9) on the control volume defined in figure 6.19. We will consider the Cartesian coordinate system shown in figure 5.2. As has been mentioned in section 6.2.7, the coordinate system $\widehat{E}$, as well as the control volume are moving in the streamwise direction at the constant dune propagation velocity $\boldsymbol{u}_D = (u_D, 0, 0)^T$. Consequently, the time rate-of-change of momentum is zero and equation (C.9) reduces to

$$\underbrace{\oint_{\mathcal{S}} \big(\langle \widehat{\boldsymbol{u}} \rangle \otimes \langle \widehat{\boldsymbol{u}} \rangle + \langle \widehat{\boldsymbol{u}}' \otimes \widehat{\boldsymbol{u}}' \rangle\big) \cdot \widehat{\boldsymbol{n}} \mathrm{d}S}_{①} = \underbrace{- \oint_{\mathcal{S}} \langle \widehat{p} \rangle \overline{\overline{\boldsymbol{I}}} \cdot \widehat{\boldsymbol{n}} \mathrm{d}S}_{②} + \underbrace{\oint_{\mathcal{S}} \langle \widehat{\overline{\overline{\tau}}} \rangle \cdot \widehat{\boldsymbol{n}} \mathrm{d}S}_{③} + \underbrace{\int_{\mathcal{V}} \langle \widehat{\boldsymbol{F}} \rangle \mathrm{d}V}_{④} \tag{C.10}$$

Additionally, the flow is statistically invariant in the spanwise direction. It follows then $\langle \widehat{w} \rangle = \langle \widehat{u}' \widehat{w}' \rangle = \langle \widehat{v}' \widehat{w}' \rangle = 0$. Thus, the streamwise and wall-normal components of equation (C.10) can be further reduce to

$$\oint_{\mathcal{S}} \big(\langle \widehat{u} \rangle \langle \widehat{\boldsymbol{u}} \rangle + \langle \widehat{u}' \widehat{\boldsymbol{u}}' \rangle\big) \cdot \widehat{\boldsymbol{n}} \mathrm{d}S = - \oint_{\mathcal{S}} \langle \widehat{p} \rangle \widehat{\boldsymbol{i}} \cdot \widehat{\boldsymbol{n}} \mathrm{d}S + \oint_{\mathcal{S}} (\langle \widehat{\overline{\overline{\tau}}} \rangle \cdot \boldsymbol{n}) \cdot \widehat{\boldsymbol{i}} \mathrm{d}S + \int_{\mathcal{V}} \langle f_x \rangle \mathrm{d}V \tag{C.11}$$

$$\oint_{\mathcal{S}} \big(\langle \widehat{v} \rangle \langle \widehat{\boldsymbol{u}} \rangle + \langle \widehat{v}' \widehat{\boldsymbol{u}}' \rangle\big) \cdot \widehat{\boldsymbol{n}} \mathrm{d}S = - \oint_{\mathcal{S}} \langle \widehat{p} \rangle \widehat{\boldsymbol{j}} \cdot \widehat{\boldsymbol{n}} \mathrm{d}S + \oint_{\mathcal{S}} (\langle \widehat{\overline{\overline{\tau}}} \rangle \cdot \boldsymbol{n}) \cdot \widehat{\boldsymbol{j}} \mathrm{d}S + \int_{\mathcal{V}} \langle f_y \rangle \mathrm{d}V \tag{C.12}$$

respectively. $\widehat{\boldsymbol{i}}$ and $\widehat{\boldsymbol{j}}$ are unit vectors aligned in the positive streamwise and wall-normal directions respectively. In the following we evaluate each of the terms at all faces of the control volume.

① streamwise component (contribution at faces AB, CD and AC):

$$-\int_{y_a}^{L_y}\left[\langle\widehat{u}\rangle^2+\langle\widehat{u}'\widehat{u}'\rangle\right]_{AB}\mathrm{d}y+\int_{y_c}^{L_y}\left[\langle\widehat{u}\rangle^2+\langle\widehat{u}'\widehat{u}'\rangle\right]_{CD}\mathrm{d}y$$
$$+\int_{x_a}^{x_c}\left[\left(\langle\widehat{u}\rangle^2+\langle\widehat{u}'\widehat{u}'\rangle\right)\tan\alpha-\langle\widehat{u}\rangle\langle\widehat{v}\rangle-\langle\widehat{u}'\widehat{v}'\rangle\right]_{AC}\mathrm{d}x \quad \text{(C.13)}$$

① wall-normal component (contribution at faces AB, CD and AC):

$$-\int_{y_a}^{L_y}[\langle\widehat{v}\rangle\langle\widehat{u}\rangle+\langle\widehat{u}'\widehat{v}'\rangle]_{AB}\mathrm{d}y+\int_{y_c}^{L_y}[\langle\widehat{v}\rangle\langle\widehat{u}\rangle+\langle\widehat{u}'\widehat{v}'\rangle]_{CD}\mathrm{d}y$$
$$+\int_{x_a}^{x_c}\left[(\langle\widehat{u}\rangle\langle\widehat{v}\rangle+\langle\widehat{u}'\widehat{v}'\rangle)\tan\alpha-\langle\widehat{v}\rangle^2-\langle\widehat{v}'\widehat{v}'\rangle\right]_{AC}\mathrm{d}x \quad \text{(C.14)}$$

② streamwise component (contribution at faces AB, CD and AC):

$$\int_{y_a}^{L_y}\left[\langle\widehat{p}\rangle\right]_{AB}\mathrm{d}y-\int_{y_c}^{L_y}\left[\langle\widehat{p}\rangle\right]_{CD}\mathrm{d}y-\int_{x_a}^{x_b}\left[\langle\widehat{p}\rangle\tan(\alpha)\right]_{AC}\mathrm{d}x \quad \text{(C.15)}$$

② wall-normal component (contribution at faces AC and BD):

$$\int_{x_a}^{x_b}\left[\langle\widehat{p}\rangle\right]_{AC}\mathrm{d}x-\int_{x_a}^{x_b}\left[\langle\widehat{p}\rangle\right]_{BD}\mathrm{d}x \quad \text{(C.16)}$$

③ streamwise component (contribution at faces AB, CD and AC):

$$-\int_{y_a}^{L_y}[2\nu\partial_{\widehat{x}}\langle\widehat{u}\rangle]_{AB}\mathrm{d}y+\int_{y_c}^{L_y}[2\nu\partial_{\widehat{x}}\langle\widehat{u}\rangle]_{CD}\mathrm{d}y+\underbrace{\int_{x_a}^{x_b}\left[2\nu\partial_{\widehat{x}}\langle\widehat{u}\rangle\tan\alpha-\nu\partial_{\widehat{y}}\langle\widehat{u}\rangle-\nu\partial_{\widehat{x}}\langle\widehat{v}\rangle\right]_{AC}\mathrm{d}x}_{\widehat{f}_{visc,x}} \quad \text{(C.17)}$$

③ wall-normal component (contribution at faces AB, CD and AC):

$$-\int_{y_a}^{L_y}\left[\nu(\partial_{\widehat{x}}\langle\widehat{v}\rangle+\partial_{\widehat{y}}\langle\widehat{u}\rangle)\right]_{AB}\mathrm{d}y+\int_{y_c}^{L_y}\left[\nu(\partial_{\widehat{x}}\langle\widehat{v}\rangle+\partial_{\widehat{y}}\langle\widehat{u}\rangle)\right]_{CD}\mathrm{d}y$$
$$+\underbrace{\int_{x_a}^{x_b}\left[\nu(\partial_{\widehat{x}}\langle\widehat{v}\rangle+\partial_{\widehat{y}}\langle\widehat{u}\rangle)\tan\alpha-2\nu\partial_{\widehat{y}}\langle\widehat{v}\rangle\right]_{AC}\mathrm{d}x}_{\widehat{f}_{visc,y}} \quad \text{(C.18)}$$

If the location of the face AC with respect to the dune geometry is appropriately chosen, the sum of the terms denoted by ④ and $\widehat{f}_{visc}$ correspond to the total boundary resistive force in the given control volume. Term ④ corresponds to the resistance as a result of the fluid-solid interaction. implicitly accounts for the total stress as a result of all particles within the control volume, whether moving as bedload, suspended load or stationary.

Appendix D

Averaging operations

Wall-parallel plane and time averaging

Let us first define an indicator function $\phi_f(\mathbf{x},t)$ for the fluid phase which tells us whether a given point with a position vector $\mathbf{x}$ lies inside $\Omega_f(t)$, the part of the computational domain Ω which is occupied by the fluid at time t, as follows:

$$\phi_f(\mathbf{x},t) = \begin{cases} 1 & \text{if} \quad \mathbf{x} \in \Omega_f(t) \\ 0 & \text{else} \end{cases} . \tag{D.1}$$

The solid-phase indicator function ϕ_p is then simply given as the complement of ϕ_f, i.e.

$$\phi_p(\mathbf{x},t) = 1 - \phi_f(\mathbf{x},t) . \tag{D.2}$$

Similar to (D.2) an individual particle indicator function $\phi_p^{(l)}$ is defined as

$$\phi_p^{(l)}(\mathbf{x},t) = \begin{cases} 1 & \text{if} \quad \mathbf{x} \in \Omega_p^{(l)}(t) \\ 0 & \text{else} \end{cases} \tag{D.3}$$

where $\Omega_p^{(l)}(t)$ is the volume occupied by the lth particle at time t.

Thus, an instantaneous discrete counter of fluid sample points in a wall-parallel plane at a given wall-distance y_j and time t^m is defined as:

$$n_{xz}(y_j,t^m) = \sum_{i=1}^{N_x} \sum_{k=1}^{N_z} \phi_f(\mathbf{x}_{ijk},t^m) , \tag{D.4}$$

where N_x and N_z are the number of grid nodes in the streamwise and spanwise directions, and $\mathbf{x}_{ijk} = (x_i,y_j,z_k)^T$ denotes a discrete grid position. Furthermore, a discrete counter of fluid sample points in a wall-parallel plane at a given wall-distance y_j, sampled over N_t number of time records, is defined as:

$$n(y_j) = \sum_{m=1}^{N_t} n_{xz}(y_j,t^m) . \tag{D.5}$$

Consequently, the ensemble average of an Eulerian quantity ξ_f of the fluid phase over wall-parallel planes, while considering only grid points located in the fluid domain, is defined as:

$$\langle\xi_f\rangle_{xz}(y_j,t^m) = \frac{1}{n_{xz}(y_j,t^m)}\sum_{i=1}^{N_x}\sum_{k=1}^{N_z}\phi_f(\mathbf{x}_{ijk},t^m)\,\xi_f(\mathbf{x}_{ijk},t^m) \tag{D.6}$$

$$\langle\xi_f\rangle(y_j) = \frac{1}{n(y_j)}\sum_{m=1}^{N_t} n_{xz}(y_j,t^m)\langle\xi_f\rangle_{xz}(y_j,t^m), \tag{D.7}$$

where the operator $\langle\cdot\rangle_{xz}$ indicates an instantaneous spatial wall-parallel plane average and $\langle\cdot\rangle$ indicates an average over space and time.

Averaging in spanwise direction and time

For the purpose of averaging a statistically stationary and two-dimensional flow field, like the phase-shifted flow velocity over the dunes, averaging is performed in the spanwise direction and in time. Thus, first a two-dimensional discrete counter of fluid sample points in the x–y plane at points x_i and y_j is defined as:

$$n_{zt}(x_i,y_j) = \sum_{k=1}^{N_z}\sum_{m=1}^{N_t}\phi_f(\mathbf{x}_{ijk},t^m), \tag{D.8}$$

Consequently, the ensemble average of an Eulerian quantity ξ_f of the fluid phase over spanwise-perpendicular planes, while considering only grid points located in the fluid domain, is defined as:

$$\langle\xi_f\rangle_{zt}(x_i,y_j) = \frac{1}{n_{zt}(x_i,y_j)}\sum_{k=1}^{N_z}\sum_{m=1}^{N_t}\phi_f(\mathbf{x}_{ijk},t^m)\,\xi_f(\mathbf{x}_{ijk},t^m), \tag{D.9}$$

Similarly, the 'particle snapshot' defined in equation (6.11) is averaged as follows:

$$\langle\xi_p\rangle_{zt}(x_i,y_j) = \frac{1}{N_zN_t - n_{zt}(x_i,y_j)}\sum_{m=1}^{N_t}\sum_{k=1}^{N_z}\sum_{l=1}^{N_p}\phi_p^{(l)}(\mathbf{x}_{ijk},t^m)\xi_p^{(l)}(t^m)\,, \tag{D.10}$$

Note that, the operators D.9 and D.10 also apply to the phase-shifted fluid and particle related quantities defined in section 6.2.1.

Instantaneous box averaging

A global counter of samples in the fluid domain can be defined analogously to (D.5), viz.

$$n_{xyz}(t) = \sum_{i=1}^{N_x}\sum_{j=1}^{N_y}\sum_{k=1}^{N_z}\phi_{\mathrm{f}}(\mathbf{x}_{ijk},t), \tag{D.11}$$

where N_y is the number of grid nodes in the wall-normal direction. The instantaneous average of a quantity ξ_f over the entire volume Ω_f occupied by the fluid is then defined as follows:

$$\langle \xi_f \rangle_{xyz}(t) = \frac{1}{n_{xyz}(t)} \sum_{i=1}^{N_x} \sum_{j=1}^{N_y} \sum_{k=1}^{N_z} \phi_f(\mathbf{x}_{ijk},t)\, \xi_f(\mathbf{x}_{ijk},t)\,. \tag{D.12}$$

Binned averages over particle-related quantities

Concerning Lagrangian quantities related to particles, the domain was decomposed into discrete wall-parallel bins of thickness Δh and averaging was performed over all those particles within each bin. Similar to (D.1), we define an indicator function $\phi_{bin}^{(j)}(y)$ which tells us whether a given wall-normal position y is located inside or outside a particular bin with index j, viz.

$$\phi_{bin}^{(j)}(y) = \begin{cases} 1 & \text{if} \quad (j-1)\Delta h \leq y < j\Delta h \\ 0 & \text{else}\,. \end{cases} \tag{D.13}$$

A sample counter for each bin, sampled over an instantaneous time t^m as well as sampled over $N_t^{(p)}$ number of available snapshots of the solid phase is defined as

$$n_{p_{xz}}^{(j)}(t^m) = \sum_{l=1}^{N_p} \phi_{bin}^{(j)}(y_p^{(l)}(t^m))\,, \tag{D.14}$$

$$n_p^{(j)} = \sum_{m=1}^{N_t^{(p)}} n_{p_{xz}}^{(j)}(t^m)\,, \tag{D.15}$$

respectively. From the sample counters we can deduce the (instantaneous) average solid volume fraction in each bin, viz.

$$\langle \phi_s \rangle_{xz}(y^{(j)},t^m) = n_{p_{xz}}^{(j)}(t^m) \frac{\pi D^3}{6 L_x L_z \Delta h}\,, \tag{D.16}$$

$$\langle \phi_s \rangle(y^{(j)}) = \frac{n_p^{(j)}}{N_t^{(p)}} \frac{\pi D^3}{6 L_x L_z \Delta h}\,. \tag{D.17}$$

Alternatively, the solid volume fraction can be deduced from the indicator function defined in (D.1) viz.

$$\langle \phi_p \rangle_{zt}(x_i,y_j) = \frac{1}{N_t^{(p)} N_z} \sum_{m=1}^{N_t^{(p)}} \sum_{k=1}^{N_z} (1 - \phi_f(\mathbf{x}_{ijk},t^m))\,, \tag{D.18}$$

$$\langle \phi_p \rangle(y_j) = \frac{1}{N_x} \sum_{k=1}^{N_x} \langle \phi_p \rangle_{zt}(x_i,y_j)\,, \tag{D.19}$$

where the operator $\langle\cdot\rangle_{zt}$ indicate averaging in the spanwise direction and time and $\langle\cdot\rangle$ indicate an average over the streamwise direction, spanwise direction and time. Finally, the binned average of a Lagrangian quantity ξ_p is defined as follows:

$$\langle\xi_p\rangle_{xz}(y^{(j)},t^m) = \frac{1}{n^{(j)}_{p_{xz}}(t^m)}\sum_{l=1}^{N_p}\phi^{(j)}_{bin}(y^{(l)}_p(t^m))\,\xi^{(l)}_p(t^m), \tag{D.20}$$

$$\langle\xi_p\rangle(y^{(j)}) = \frac{1}{n^{(j)}_p}\sum_{m=1}^{N^{(p)}_t} n^{(j)}_{p_{xz}}(t^m)\langle\xi_p\rangle_{xz}(y^{(j)},t^m), \tag{D.21}$$

supposing that a finite number of samples has been encountered ($n^{(j)}_{p_{xz}}(t^m) > 0$, $n^{(j)}_p > 0$). A bin thickness of $\Delta h = D/4$ was adopted to evaluate the bin-averages unless otherwise stated.

Bibliography

Allen, J. (1968). *Current ripples: their relation to patterns of water and sediment motion*. North-Holland Pub. Co. (cited on p. 2).

Andreotti, B. and P. Claudin (2013). "Aeolian and subaqueous bedforms in shear flows." *Philos. Trans. A. Math. Phys. Eng. Sci.* 371(2004), p. 20120364 (cited on pp. 2, 105, 108).

Andreotti, B., P. Claudin, and S. Douady (2002). "Selection of dune shapes and velocities Part 1: Dynamics of sand, wind and barchans". *Eur. Phys. J. B - Condens. Matter*, 28(3), pp. 321–339 (cited on p. 2).

ASCE Task Committee (2002). "Flow and Transport over Dunes". *J. Hydraul. Eng.* 128(8), pp. 726–728 (cited on p. 3).

Aussillous, P., J. Chauchat, M. Pailha, M. Médale, and É. Guazzelli (2013). "Investigation of the mobile granular layer in bedload transport by laminar shearing flows". *J. Fluid Mech.* 736, pp. 594–615 (cited on pp. 10, 29, 33–37, 39–44, 48–50, 83, 112).

Bagnold, R. A. (1941). *The Physics of Blown Sand and Desert Dunes*. Chapman and Hall (cited on p. 2).

Bagnold, R. A. (1956). "The flow of cohesionless grains in fluids". *Philos. Trans. R. Soc. London. Ser. A*, 249(964), pp. 235–297 (cited on p. 2).

Batchelor, G. K. (1967). *An Introduction to Fluid Dynamics*. Cambridge University Press (cited on p. 5).

Bennett, S. J. and J. L. Best (1995). "Mean flow and turbulence structure over fixed, two-dimensional dunes: implications for sediment transport and bedform stability". *Sedimentology*, 42, pp. 491–513 (cited on pp. 14, 94).

Best, J. (2005). "The fluid dynamics of river dunes: A review and some future research directions". *J. Geophys. Res.* 110(F4), F04S02 (cited on pp. 1–3, 13, 14, 62, 84, 86, 96, 107).

Betat, A., C. A. Kruelle, V. Frette, and I. Rehberg (2002). "Long-time behavior of sand ripples induced by water shear flow." *Eur. Phys. J. E. Soft Matter*, 8(5), pp. 465–76 (cited on pp. 12, 13, 59, 64, 73).

Blondeaux, P. (1990). "Sand ripples under sea waves Part 1. Ripple formation". *J. Fluid Mech.* 218, pp. 1–17 (cited on p. 2).

Blondeaux, P., E. Foti, and G. Vittori (2015). "A theoretical model of asymmetric wave ripples". *Philos. Trans. R. Soc. London. Ser. A*, 373, p. 20140112 (cited on pp. 2, 62).

Blondeaux, P. (2001). "M ECHANICS OF C OASTAL F ORMS". *Annu. Rev. Fluid Mech.* 33(1), pp. 339–370 (cited on p. 2).

Boyer, F., É. Guazzelli, and O. Pouliquen (2011). "Unifying Suspension and Granular Rheology". *Phys. Rev. Lett.* 107(18), p. 188301 (cited on pp. viii, 34, 47–50, 112).

Brändle de Motta, J. C., W.-P. Breugem, B. Gazanion, J.-L. Estivalezes, S. Vincent, and E. Climent (2013). "Numerical modelling of finite-size particle collisions in a viscous fluid". *Phys. Fluids*, 25(8), p. 083302 (cited on p. 22).

Bridge, J. S. and J. L. Best (1988). "Flow, sediment transport and bedform dynamics over the transition from dunes to upper-stage plane beds: implications for the formation of planar laminae". *Sedimentology*, 35, pp. 753–763 (cited on p. 3).

Cate, A. ten, C. H. Nieuwstad, J. J. Derksen, and H. E. A. Van den Akker (2002). "Particle imaging velocimetry experiments and lattice-Boltzmann simulations on a single sphere settling under gravity". *Phys. Fluids*, 14(11), p. 4012 (cited on p. 23).

Cellino, M. and W. H. Graf (2000). "Experiments on suspension flow in open channels with bedforms". *J. Hydraul. Res.* 38(4), pp. 289–298 (cited on p. 96).

Chan-Braun, C., M. García-Villalba, and M. Uhlmann (2011). "Force and torque acting on particles in a transitionally rough open-channel flow". *J. Fluid Mech.* 684, pp. 441–474 (cited on pp. 18, 79, 80).

Chan-Braun, C., M. García-Villalba, and M. Uhlmann (2013). "Spatial and temporal scales of force and torque acting on wall-mounted spherical particles in open channel flow". *Phys. Fluids*, 25(7), p. 075103 (cited on p. 18).

Chan-Braun, C. (2012). "Turbulent open channel flow, sediment erosion and sediment transport". Phd thesis. Institute for Hydromechanics, Karlsruhe Institute of Technology (KIT) (cited on pp. 98, 108).

Charru, F. (2006). "Selection of the ripple length on a granular bed sheared by a liquid flow". *Phys. Fluids*, 18(12), p. 121508 (cited on pp. 11, 12, 68, 105).

Charru, F., B. Andreotti, and P. Claudin (2013). "Sand Ripples and Dunes". *Annu. Rev. Fluid Mech.* 45(1), pp. 469–493 (cited on pp. ix, 2, 9, 62, 66, 76, 105, 106, 108, 109, 113).

Charru, F. and E. M. Franklin (2012). "Subaqueous barchan dunes in turbulent shear flow. Part 2. Fluid flow". *J. Fluid Mech.* 694(Best 2005), pp. 131–154 (cited on pp. 14, 86, 87, 93, 96, 97, 102, 105, 107).

Charru, F. and E. J. Hinch (2006a). "Ripple formation on a particle bed sheared by a viscous liquid. Part 1. Steady flow". *J. Fluid Mech.* 550(-1), pp. 111–121 (cited on pp. 11–13).

Charru, F. and E. J. Hinch (2006b). "Ripple formation on a particle bed sheared by a viscous liquid. Part 2. Oscillating flow". *J. Fluid Mech.* 550(-1), pp. 123–137 (cited on p. 2).

Charru, F., E. Larrieu, J.-B. Dupont, and R. Zenit (2007). "Motion of a particle near a rough wall in a viscous shear flow". *J. Fluid Mech.* 570, p. 431 (cited on p. 10).

Charru, F., H. Mouilleron, and O. Eiff (2004). "Erosion and deposition of particles on a bed sheared by a viscous flow". *J. Fluid Mech.* 519, pp. 55–80 (cited on pp. 10, 11, 36).

Charru, F. and H. Mouilleron-Arnould (2002). "Instability of a bed of particles sheared by a viscous flow". *J. Fluid Mech.* 452, pp. 303–323 (cited on p. 11).

Claudin, P., F. Charru, and B. Andreotti (2011). "Transport relaxation time and length scales in turbulent suspensions". *J. Fluid Mech.* 671, pp. 491–506 (cited on p. 105).

Cleary, P. W. and M. Prakash (2004). "Discrete-element modelling and smoothed particle hydrodynamics: potential in the environmental sciences." *Philos. Trans. A. Math. Phys. Eng. Sci.* 362(1822), pp. 2003–30 (cited on p. 22).

Coleman, S. E., J. J. Fedele, and M. H. Garcia (2003). "Closed-Conduit Bed-Form Initiation and Development". *J. Hydraul. Eng.* 129(December), pp. 956–965 (cited on pp. 12, 64, 73).

Coleman, S. E. and B. W. Melville (1994). "Bed-form development". *J. Hydraul. Eng.* 120(4), pp. 544–560 (cited on pp. 3, 12).

Coleman, S. E. and V. I. Nikora (2009). "Bed and flow dynamics leading to sediment-wave initiation". *Water Resour. Res.* 45(4) (cited on pp. 11, 59, 69).

Coleman, S. E. and V. I. Nikora (2011). "Fluvial dunes: initiation, characterization, flow structure". *Earth Surf. Process. Landforms*, 36(1), pp. 39–57 (cited on pp. 1, 59, 63, 86).

Colombini, M. (2004). "Revisiting the linear theory of sand dune formation". *J. Fluid Mech.* 502, pp. 1–16 (cited on pp. 11, 12).

Colombini, M. and a. Stocchino (2008). "Finite-amplitude river dunes". *J. Fluid Mech.* 611, pp. 283–306 (cited on pp. 13, 62).

Colombini, M. and a. Stocchino (2011). "Ripple and dune formation in rivers". *J. Fluid Mech.* 673, pp. 121–131 (cited on pp. 11, 12).

Colombini, M. (2014). "A decade's investigation of the stability of erodible stream beds". *J. Fluid Mech.* 756, pp. 1–4 (cited on pp. 9, 52, 111).

Constantinides, G., C. Tweedie, D. Holbrook, P. Barragan, J. Smith, and K. V. Vliet (2008). "Quantifying deformation and energy dissipation of polymeric surfaces under localized impact". *Mat. Sci. Eng. A*, 489(1-2), pp. 403–412 (cited on p. 31).

Crowe, C. and C. Sommerfeld (1998). *Multiphase flows with droplets and particles.* CRC press (cited on pp. 18, 19, 21).

Cundall, P. A. and O. D. L. Strack (1979). "A discrete numerical model for granular assemblies". *Géotechnique*, 29(1), pp. 47–65 (cited on p. 19).

Dey, S. (2014). *Fluvial Hydrodynamics.* Springer Berlin Heidelberg (cited on p. 10).

Endo, N., T. Sunamura, and H. Takimoto (2005). "Barchan ripples under unidirectional water flows in the laboratory: formation and planar morphology". *Earth Surf. Process. Landforms*, 30(13), pp. 1675–1682 (cited on p. 2).

Engelund, F. and J. Fredsoe (1982). "Sediment Ripples and Dunes". *Annu. Rev. Fluid Mech.* 14(1), pp. 13–37 (cited on pp. 3, 4, 94).

Exner, F. M. (1925). "Uber die Wechselwirkung zwischen Wasser und Geschiebe in Flüssen". *Akad. der Wiss Wien, Math-Naturwissenschafliche Klasse, Sitzungsberichte, Abt IIa*, 134(165-203) (cited on p. 9).

Foerster, S. F., M. Y. Louge, H. Chang, and K. Allia (1994). "Measurements of the collision properties of small spheres". *Phys. Fluids*, 6(3), p. 1108 (cited on pp. 19, 31).

Fourrière, A., P. Claudin, and B. Andreotti (2010). "Bedforms in a turbulent stream: formation of ripples by primary linear instability and of dunes by nonlinear pattern coarsening". *J. Fluid Mech.* 649, pp. 287–328 (cited on pp. 12, 13, 105).

Franklin, E. M. and F. Charru (2011). "Subaqueous barchan dunes in turbulent shear flow. Part 1. Dune motion". *J. Fluid Mech.* 675(1988), pp. 199–222 (cited on pp. 12, 53, 68).

Franklin, E. and F. Charru (2009). "Morphology and displacement of dunes in a closed-conduit flow". *Powder Technol.* 190(1-2), pp. 247–251 (cited on p. 2).

Fredsøe, J. (1974). "On the development of dunes in erodible channels". *J. Fluid Mech.* 64(01), pp. 1–16 (cited on p. 11).

Führböter, A. (1983). "Zur Bildung von makroskopischen Ordnungsstructuren (Strömungsriffel und Dünen) aus sehr kleinen zufallstörungen". Mitteilungen. Technischen Universität Braunschweig (cited on p. 73).

García, M. H., ed. (2008a). *Sedimentation engineering: processes, measurements, modeling, and practice.* 110th ed. ASCE Publications (cited on p. 10).

García, M. H., ed. (2008b). *Sedimentation Engineering. Processes, Measurments, Modeling and Practice.* American Society of Civil Engineers (cited on pp. 1, 3, 10).

García-Villalba, M., A. G. Kidanemariam, and M. Uhlmann (2012). "DNS of vertical plane channel flow with finite-size particles: Voronoi analysis, acceleration statistics and particle-conditioned averaging". *Int. J. Multiph. Flow*, 46, pp. 54–74 (cited on pp. 18, 45).

Glowinski, R., R. Pan, T. I. Hesla, and D. D. Joseph (1999). "A distributed Lagrange multiplier/fictitious domain method for particulate flows". *Int. J. Multiph. Flow*, 25, pp. 755–794 (cited on p. 19).

Goharzadeh, A., A. Khalili, and B. B. Jørgensen (2005). "Transition layer thickness at a fluid-porous interface". *Phys. Fluids*, 17(5), p. 057102 (cited on p. 44).

Gondret, P., M. Lance, and L. Petit (2002). "Bouncing motion of spherical particles in fluids". *Phys. Fluids*, 14(2), pp. 643–652 (cited on pp. 23, 25–27).

Groh, C., A. Wierschem, N. Aksel, I. Rehberg, and C. Kruelle (2008). "Barchan dunes in two dimensions: Experimental tests for minimal models". *Phys. Rev. E*, 78(2), p. 021304 (cited on p. 2).

Hong, A., M. Tao, and A. Kudrolli (2015). "Onset of erosion of a granular bed in a channel driven by fluid flow". *Phys. Fluids*, 27, p. 013301 (cited on p. 44).

Hoomans, B., J. Kuipers, W. Briels, and W. van Swaaij (1996). "Discrete particle simulation of bubble and slug formation in a two-dimensional gas-fluidised bed: A hard-sphere approach". *Chem. Eng. Sci.* 51(1), pp. 99–118 (cited on p. 19).

Jackson, R. (2000). *The dynamics of fluidized particles*. Cambridge University Press (cited on p. 7).

Jeong, J. and F. Hussain (1995). "On the identification of a vortex". *J. Fluid Mech.* 285, pp. 69–94 (cited on p. 85).

Jiménez, J. (2004). "Turbulent flows over rough walls". *Annu. Rev. Fluid Mech.* 36, pp. 173–196 (cited on pp. 80, 107).

Jiménez, J. and P. Moin (1991). "The minimal flow unit in near-wall turbulence". *J. Fluid Mech.* 225, pp. 213–240 (cited on pp. 68, 69, 74, 112).

Joseph, G. G. and M. L. Hunt (2004). "Oblique particlewall collisions in a liquid". *J. Fluid Mech.* 510, pp. 71–93 (cited on pp. 23, 31).

Joseph, G. G., R. Zenit, M. L. Hunt, and A. M. Rosenwinkel (2001). "Particle-wall collisions in a viscous fluid". *J. Fluid Mech.* 433, pp. 329–346 (cited on pp. 23, 25, 27).

Julien, P. (1998). *Erosion and Sedimentation*. Cambridge University Press (cited on pp. 3, 4).

Kempe, T. and J. Fröhlich (2012). "Collision modelling for the interface-resolved simulation of spherical particles in viscous fluids". *J. Fluid Mech.* 709, pp. 445–489 (cited on p. 22).

Kennedy, J. F. (1963). "The mechanics of dunes and antidunes in erodible-bed channels". *J. Fluid Mech.* 16(4), pp. 521–544 (cited on p. 11).

Khosronejad, A. and F. Sotiropoulos (2014). "Numerical simulation of sand waves in a turbulent open channel flow". *J. Fluid Mech.* 753, pp. 150–216 (cited on p. 14).

Kidanemariam, A. G. (2010). "Numerical simulation of sediment transport in an open channel flow with fully resolved particles". Master thesis. Institute for Hydromechanics, Karlsruhe Institute of Technology (KIT) (cited on p. 98).

Kidanemariam, A. G., C. Chan-Braun, T. Doychev, and M. Uhlmann (2013). "Direct numerical simulation of horizontal open channel flow with finite-size, heavy particles at low solid volume fraction". *New J. Phys.* 15(2), p. 025031 (cited on pp. 18, 45, 46, 55, 76, 77, 84, 91, 93, 98, 100, 108).

Kidanemariam, A. G. and M. Uhlmann (2014a). "Direct numerical simulation of pattern formation in subaqueous sediment". *J. Fluid Mech.* 750, R2 (cited on pp. 52, 62).

Kidanemariam, A. G. and M. Uhlmann (2014b). "Interface-resolved direct numerical simulation of the erosion of a sediment bed sheared by laminar channel flow". *Int. J. Multiph. Flow*, pp. 1–27 (cited on p. 29).

Langlois, V. and A. Valance (2007). "Initiation and evolution of current ripples on a flat sand bed under turbulent water flow." *Eur. Phys. J. E. Soft Matter*, 22(3), pp. 201–8 (cited on pp. 3, 11, 12, 51, 52, 59–62, 73).

Leighton, D. and A. Acrivos (1986). "Viscous resuspension". *Chem. Eng. Sci.* 41(6), pp. 1377–1384 (cited on pp. 11, 39).

Li, X., M. L. Hunt, and T. Colonius (2011). "A contact model for normal immersed collisions between a particle and a wall". *J. Fluid Mech.* 691, pp. 123–145 (cited on p. 22).

Lobkovsky, A. E., A. V. Orpe, R. Molloy, A. Kudrolli, and D. H. Rothman (2008). "Erosion of a granular bed driven by laminar fluid flow". *J. Fluid Mech.* 605, pp. 47–58 (cited on pp. 10, 33, 34, 36).

Loiseleux, T., P. Gondret, M. Rabaud, and D. Doppler (2005). "Onset of erosion and avalanche for an inclined granular bed sheared by a continuous laminar flow". *Phys. Fluids*, 17(10), p. 103304 (cited on pp. 10, 11, 36).

Maddux, T. B., S. R. Mclean, and J. M. Nelson (2003). "Turbulent flow over three-dimensional dunes: 2. Fluid and bed stresses". *J. Geophys. Res.* 108, p. 6010 (cited on p. 102).

McLean, B. S. R. and J. D. Smith (1986). "A model for flow over two-dimensional bed forms". *J. Hydraul. Eng.* 112(4), pp. 300–317 (cited on pp. 14, 94, 102).

McLean, S. R., J. M. Nelson, and S. R. Wolfe (1994). "Turbulence structure over two-dimensional bed forms: Implications for sediment transport". *J. Geophys. Res.* 99, pp. 12729–12747 (cited on pp. 14, 84, 86, 94, 106).

Meyer-Peter, E. and R. Müller (1948). "Formulas for bed-load transport". In: *Proc. 2nd Meet.* Ed. by IAHR. Stockholm, Sweden, pp. 39–64 (cited on p. 83).

Mittal, R. and G. Iaccarino (2005). "Immersed boundary methods". *Annu. Rev. Fluid Mech.* 37, pp. 239–261 (cited on pp. 17, 18).

Mouilleron, H., F. Charru, and O. Eiff (2009). "Inside the moving layer of a sheared granular bed". *J. Fluid Mech.* 628, p. 229 (cited on pp. 10, 39, 44, 50).

Nelson, J. M., S. R. McLean, and S. R. Wolfe (1993). *Mean flow and turbulence fields over two-dimensional bed forms* (cited on pp. 14, 94).

Nelson, J. M. and J. D. Smith (1989). "Mechanics of flow over ripples and dunes". *J. Geophys. Res.* 94, pp. 8146–8162 (cited on pp. 14, 94).

Nikora, V. I., A. N. Sukhodolov, and P. M. Rowinski (1997). "Statistical sand wave dynamics in one-directional water flows". *J. Fluid Mech.* 351, pp. 17–19 (cited on p. 63).

Nikora, V., I. McEwan, S. McLean, S. Coleman, D. Pokrajac, and R. Walters (2007a). "Double-Averaging Concept for Rough-Bed Open-Channel and Overland Flows: Theoretical Background". *J. Hydraul. Eng.* 133(8), pp. 873–883 (cited on pp. 86, 89).

Nikora, V., S. Mclean, S. Coleman, D. Pokrajac, I. Mcewan, L. Campbell, J. Aberle, D. Clunie, and K. Koll (2007b). "Double-Averaging Concept for Rough-Bed Open-Channel and Overland Flows: Applications". *J. Hydraul. Eng.* 133, pp. 884–895 (cited on pp. 101, 102, 108).

Omidyeganeh, M. and U. Piomelli (2011). "Large-eddy simulation of two-dimensional dunes in a steady, unidirectional flow". *J. Turbul.* 12(April 2014), N42 (cited on pp. 14, 84, 94).

Ouriemi, M., P. Aussillous, and É. Guazzelli (2009a). "Sediment dynamics. Part 2. Dune formation in pipe flow". *J. Fluid Mech.* 636, pp. 295–319 (cited on pp. 3, 11–13, 15, 51, 52, 59, 60, 62, 64, 71, 73, 86, 112).

Ouriemi, M., P. Aussillous, and É. Guazzelli (2009b). "Sediment dynamics. Part 1. Bed-load transport by laminar shearing flows". *J. Fluid Mech.* 636, pp. 295–319 (cited on pp. 10, 11, 36, 37, 39, 44, 50, 81).

Ouriemi, M., P. Aussillous, M. Medale, Y. Peysson, and É. Guazzelli (2007). "Determination of the critical Shields number for particle erosion in laminar flow". *Phys. Fluids*, 19(6), p. 061706 (cited on pp. 10, 36, 37).

Peskin, C. S. (2002). "The immersed boundary method". *Acta Numerica*, 11 (cited on p. 18).

Pope, S. B. (2000). *Turbulent flows.* 1st. Cambrigde: Cambridge University Press (cited on pp. 5–8, 17, 41, 80, 102, 107, 108, 125).

Prosperetti, A. and G. Tryggvason, eds. (2007). *Computational Methods for Multiphase Flow.* Cambridge University Press (cited on p. 17).

Raudkivi, A. J. (1997). "Ripples on Stream Bed". *J. Hydraul. Eng.* 123(1), pp. 58–64 (cited on p. 11).

Richards, K. J. (1980). "The formation of ripples and dunes on an erodible bed". *J. Fluid Mech.* 99(3), pp. 597–618 (cited on pp. 11, 12).

Righetti, M. and G. P. Romano (2004). "Particle-fluid interactions in a plane near-wall turbulent flow". *J. Fluid Mech.* 505, pp. 93–121 (cited on p. 108).

Rousseaux, G., A. Stegner, and J. E. Wesfreid (2004). "Wavelength selection of rolling-grain ripples in the laboratory". *Phys. Rev. E*, 69 (3), p. 031307 (cited on p. 2).

Sauermann, G., K. Kroy, and H. Herrmann (2001). "Continuum saltation model for sand dunes". *Phys. Rev. E*, 64(3), p. 031305 (cited on pp. 2, 12).

Scherer, M. A., F. Melo, M. Marder, and I. Introduction (1999). "Sand ripples in an oscillating annular sand–water cell". *Phys. Fluids*, 11(1), pp. 58–67 (cited on p. 2).

Schlichting, H. and K. Gersten (2000). *Boundary layer theory.* 8th Edition. Springer (cited on p. 108).

Seminara, G. (2010). "Fluvial Sedimentary Patterns". *Annu. Rev. Fluid Mech.* 42(1), pp. 43–66 (cited on pp. 2, 9, 62).

Shields, A. (1936). "Anwendung der Aehnlichkeitsmechanik und der Turbulenzforschung auf die Geschiebebewegung". Mitteilungen der Versuchanstalt für Wasserbau und Schiffbau. Technischen Hochschule Berlin (cited on p. 9).

Silbert, L., D. Ertas, G. Grest, T. Halsey, D. Levine, and S. Plimpton (2001). "Granular flow down an inclined plane: Bagnold scaling and rheology". *Phys. Rev. E*, 64(5), p. 051302 (cited on p. 30).

Simeonov, J. a. and J. Calantoni (2012). "Modeling mechanical contact and lubrication in Direct Numerical Simulations of colliding particles". *Int. J. Multiph. Flow*, 46, pp. 38–53 (cited on p. 22).

Sleath, J. F. A. (1976). "On rolling-grain ripples". *J. of Hydraul. Res.* 14(1), pp. 69–81 (cited on p. 2).

Stoesser, T., C. Braun, M. García-Villalba, and W. Rodi (2008). "Turbulence Structures in Flow over Two-Dimensional Dunes". *J. Hydraul. Eng.* 134(1), pp. 42–55 (cited on pp. 14, 84, 94).

Sumer, B. M. and M. Bakioglu (1984). "On the formation of ripples on an erodible bed". *J. Fluid Mech.* 144, pp. 177–190 (cited on pp. 11, 12).

Tanaka, T. and Y. Tsuji (1991). "Numerical simulation of gas-solid two-phase flow in a vertical pipe: on the effect of inter-particle collision". In: *4th Int. Symp. Gas-Solid Flows.* Pp. 123–128 (cited on p. 19).

Tsuji, Y., T. Kawaguchi, and T. Tanaka (1993). "Discrete particle simulation of two-dimensional fluidized bed". *Powder Technol.* 77, pp. 79–87 (cited on pp. 19, 31).

Uhlmann, M. (2004). *New Results on the Simulation of Particulate Flows.* Tech. rep. Madrid: CIEMAT, pp. 1–38 (cited on p. 18).

Uhlmann, M. (2005a). "An immersed boundary method with direct forcing for the simulation of particulate flows". *J. Comput. Phys.* 209(2), pp. 448–476 (cited on pp. 18, 111, 119, 123).

Uhlmann, M. (2008). "Interface-resolved direct numerical simulation of vertical particulate channel flow in the turbulent regime". *Phys. Fluids*, 20(5), p. 053305 (cited on pp. 18, 47, 77, 100, 101, 108).

Uhlmann, M. and J. Dušek (2014). "The motion of a single heavy sphere in ambient fluid: a benchmark for interface-resolved particulate flow simulations with significant relative velocities". *Int. J. Multiphase Flow*, 59, pp. 221–243 (cited on p. 18).

Uhlmann, M. (2005b). "An improved fluid-solid coupling method for DNS of particulate flow on a fixed mesh". In: *Proc. 11th Work. Two-Phase Flow Predict.* Ed. by M. Sommerfel. Merseburg, Germany: Universität Halle (cited on p. 18).

Uhlmann, M. (2006). "Experience with DNS of particulate flow using a variant of the immersed boundary method". In: *Proc. ECCOMAS CFD 2006.* Ed. by P. Wesseling, E. Onate, and J. Périaux. Egmond aan Zee, The Netherlands: TU Delft (cited on p. 18).

Valance, a. (2005). "Formation of ripples over a sand bed submitted to a turbulent shear flow". *Eur. Phys. J. B*, 45(3), pp. 433–442 (cited on p. 12).

Valance, a. and V. Langlois (2005). "Ripple formation over a sand bed submitted to a laminar shear flow". *Eur. Phys. J. B*, 43(2), pp. 283–294 (cited on p. 12).

Verzicco, R. and P. Orlandi (1996). "A Finite-Difference Scheme for Three-Dimensional Incompressible Flows in Cylindrical Coordinates". *J. Comput. Phys.* 123(2), pp. 402–414 (cited on p. 18).

Vittori, G. and P. Blondeaux (1990). *Sand ripples under sea waves Part 2. Finite-amplitude development* (cited on p. 2).

Wachs, A. (2009). "A DEM-DLM/FD method for direct numerical simulation of particulate flows: Sedimentation of polygonal isometric particles in a Newtonian fluid with collisions". *Comput. Fluids*, 38(8), pp. 1608–1628 (cited on pp. 19, 22, 119).

Wong, M. and G. Parker (2006). "Reanalysis and Correction of Bed-Load Relation of Meyer-Peter and Müller Using Their Own Database". *J. Hydraul. Eng.* 132, pp. 1159–1168 (cited on pp. viii, 53, 83, 107, 113).

Yalin, M. S. (1977). *Mechanics of sediment transport.* 2nd. Pergamon press (cited on pp. 1–3, 77, 86, 102).

Yang, F.-L. and M. L. Hunt (2006). "Dynamics of particle-particle collisions in a viscous liquid". *Phys. Fluids*, 18(12), p. 121506 (cited on p. 23).

Yang, F.-L. and M. L. Hunt (2008). "A mixed contact model for an immersed collision between two solid surfaces." *Philos. Trans. A. Math. Phys. Eng. Sci.* 366(1873), pp. 2205–18 (cited on p. 22).

Yuan, J. and U. Piomelli (2014). "Roughness effects on the Reynolds stress budgets in near-wall turbulence". *J. Fluid Mech.* 760, R1 (cited on p. 80).

List of Figures

List of Tables

Dissertationsreihe am Institut für Hydromechanik
Karlsruher Institut für Technologie (KIT)

ISSN 1439-4111

Die Bände sind unter www.ksp.kit.edu als PDF frei verfügbar oder als Druckausgabe bestellbar.

2004/1 **Volker Weitbrecht**
Influence of dead-water zones on the dispersive mass-transport in rivers. 2004
ISBN 3-937300-07-4

2004/2 **Christof Hinterberger**
Dreidimensionale und tiefengemittelte Large-Eddy-Simulation von Flachwasserströmungen. 2004
ISBN 3-937300-18-X

2005/1 **Cornelia Frank**
Wirksamkeit von dünnen Windschutzstreifen auf Sockelwällen in luv- und leeseitiger Anordnung. 2005
ISBN 3-937300-55-4

2005/2 **Carl Friedrich von Carmer**
Shallow turbulent wake flows: momentum and mass transfer due to large-scale coherent vortical structures. 2005
ISBN 3-937300-63-5

2005/3 **Muhammad Ikhwan**
Investigation of flow and pressure characteristics around pyramidal buildings. 2005
ISBN 3-937300-72-4

2005/4 **Herlina**
Gas transfer at the air-water interface in a turbulent flow environment. 2005
ISBN 3-937300-74-0

2006/1 **Manual García-Villalba Navaridas**
Large eddy simulation of turbulent swirling jets. 2006
ISBN 3-86644-015-4

2006/2 **Klaas E. Heinrich**
Hydraulische Steuerung einer Wasser-Alkohol-Zirkulation mittels Grundwasser-Zirkulations-Brunnen. 2006
ISBN 3-86644-055-3

2006/3 **Tobias Bleninger**
Coupled 3D hydrodynamic models for submarine outfalls. Denvironmental hydraulic design and control of multiport diffusers. 2006
ISBN 978-3-86644-160-6

2006/3 **Patrick Heneka**
Schäden durch Winterstürme – das Schadensrisiko von Wohngebäuden in Baden-Württemberg. 2006
ISBN 978-3-86644-117-0

2007/1 **Maria Eletta Negretti**
Hydrodynamic instabilities and entrainment in two-layer stratified flows down a slope. 2007
ISBN 978-3-86644-170-5

2007/2 **Gregor Kühn**
Untersuchungen zur Feinsedimentdynamik unter Turbulenzeinfluss. 2007
ISBN 978-3-86644-205-4

2008/1 **Martin Detert**
Hydrodynamic processes at the water-sediment interface of streambeds. 2008
ISBN 978-3-86644-244-3

2008/2 **Christof Gromke**
Einfluss von Bäumen auf die Durchlüftung von innerstädtischen Straßenschluchten. 2008
ISBN 978-3-86644-339-6

2009/1 **Ekkehart Bethge**
Risikoberechnung zum Schadstoffeintrag aus Hochwasserretentionsräumen in einen Grundwasserleiter. 2009
ISBN 978-3-86644-426-3

2012/1 **Clemens Chan-Braun**
Turbulent open channel flow, sediment erosion and sediment transport. 2012
ISBN 978-3-86644-900-8

2015/1 **Todor Doychev**
The dynamics of finite-size settling particles. 2015
ISBN 978-3-7315-0307-1

2016/1 **Aman Ghebremichael Kidanemariam**
The formation of patterns in subaqueous sediment. 2016
ISBN 978-3-7315-0532-7